Kawthar BELKAALOUL

Lactic acid bacteria and intestinal function

Kawthar BELKAALOUL

Lactic acid bacteria and intestinal function

Experimentation and application

ScienciaScripts

Imprint

Any brand names and product names mentioned in this book are subject to trademark, brand or patent protection and are trademarks or registered trademarks of their respective holders. The use of brand names, product names, common names, trade names, product descriptions etc. even without a particular marking in this work is in no way to be construed to mean that such names may be regarded as unrestricted in respect of trademark and brand protection legislation and could thus be used by anyone.

Cover image: www.ingimage.com

This book is a translation from the original published under ISBN 978-620-6-70600-7.

Publisher:
Sciencia Scripts
is a trademark of
Dodo Books Indian Ocean Ltd. and OmniScriptum S.R.L publishing group

120 High Road, East Finchley, London, N2 9ED, United Kingdom
Str. Armeneasca 28/1, office 1, Chisinau MD-2012, Republic of Moldova, Europe
Printed at: see last page
ISBN: 978-620-8-07077-9

Contents

Summary 2

Introduction..3

CHAPTER 1 4

CHAPTER 2 8

CHAPTER 3 14

CHAPTER 4 19

CHAPTER 5 23

CHAPTER 6 34

Conclusion 71

References 72

Summary

Fermented milks are sources of bioactive peptides that can modulate various bodily functions such as the digestive system, the cardiovascular system and the immune system. Five lactic acid bacteria were studied in this study: *Lactobacillus paracasei, Enterococcus faecalis* DAPTO 512, *Enterococcus faecium,* and two co-cultures (*bifidobacterium longum - lactobacillus plantarum*) and (*bifidobacterium longum.-streptococcus thermophillus*), isolated locally and identified from cow's milk. The performance of the selected strains during fermentation and their probiotic potential were evaluated *in vitro.* The hydrolysates obtained after fermentation were characterised by electrophoresis (SDS-PAGE) and chromatography (HPLC). The antioxidant capacity of the hydrophobic peptide fractions was measured by ABTS. The preventive and therapeutic effects of the different hydrolysates were evaluated *ex-vivo* in a Ussing chamber on the intestinal mucosa of Balb/c mice sensitised to ß-Lg by the intraperitoneal route and orally to bovine milk. A histological study was carried out to assess the state of the intestinal architecture. The results show that the selected strains have good fermentation performance, their proteolytic activity is high and they hydrolyse sodium caseinates and ß-Lg to varying degrees. The chromatographic profile of the peptides generated shows that they appear in the hydrophobic fraction of the mobile phase. Assessment of the antioxidant capacity of these peptide fractions showed that the peptides produced by the *L.paracasei* strain have a marked antioxidant activity. Evaluation of the preventive effect showed that animals given the hydrolysates produced slightly lower levels of serum anti-Blg IgG and IgE than those given the positive control. The Ussing chamber study showed that the response to allergen stimulation was significantly lower than that observed in the positive control group. The study of the therapeutic effect of the different hydrolysates also showed that the lgE rate remained comparable to that of the positive control group. On the other hand, the Ussing chamber allergy challenge tests showed a reduction in short circuit current (Isc) and conductance (G). Histological studies showed that the groups that received hydrolysates had minor histological damage with less inflammatory signs and villi that were more or less enlarged. The hydrolysates produced by the selected strains have a preventive and therapeutic effect in our experimental model and also appear to have a protective effect on the intestinal epithelium. The *L. paracasei* strain is an interesting candidate since the therapeutic and preventive effects are significantly more marked.

Key words : lactic acid bacteria- fermented milks- proteolysis- antioxidant activity- immunomodulation- bioactive peptides- Ussing chamber- Balb/c mice.

Introduction

Fermented milks have been consumed for thousands of years because of their pleasant taste and good shelf life.

Lactic acid bacteria are Gram-positive bacteria that produce lactic acid as the main product of their metabolism. They comprise 12 bacterial genera, the most extensively studied of which are *Lactobacillus*, *Lactococcus*, *Streptococcus*, *Enterococcus*, *Pediococcus* and *Bifidobacterium* (Stackebrandt and Teuber, 1988).

These micro-organisms are used in various processes in the food industry. They are recognised as safe by the health authorities (Jankovic et *al.*, 2010). They play an important role in biochemical reactions during milk acidification, flavour development and proteolysis. Proteolysis is considered to be one of the most important biochemical processes involved in the manufacture of food industry products. The ability to secrete extracellular proteinases is very important for the growth of lactic acid bacteria. They hydrolyse milk proteins, providing essential amino acids for growth (Fira et *al.*, 2001). Proteolysis during bacterial fermentation improves the digestibility and nutritional quality of the final dairy product, as well as the change in texture and flavour (Fira et *al.*, 2001).

But nowadays their popularity is more associated with their positive effects on health. Milk proteins possess numerous biological activities that make these components effective at improving human health (Hernández-Ledesma et *al.*, 2014). In recent years, it has been recognised that not only intact caseins and whey proteins have a nutritional effect, the bioactive peptides derived from these fractions possess physiological properties (Clare & Swaisgood, 2000; Haquea et *al.*, 2009). Numerous biological properties have been described in the literature, including peptides with antimicrobial (López-Expósito & Recio, 2006), cholesterol-lowering (Hartmann & Meisel, 2007), antihypertensive (Korhonen, 2009; Jäkälä and Vapaatalo, 2010; Moslehishad et *al.*, 2013), opioid, immunomodulatory and antioxidant activity (Moslehishad et *al.*, 2013).

Bioactive peptides have recently been used as nutraceuticals. The use of dairy protein hydrolysates after lactic fermentation in cow's milk protein allergy is an interesting avenue.

Food allergies are on the rise worldwide. They affect 2.5% of children under the age of 3 (Chatchatee et *al.*, 2001; Ross et *al.*, 2005). Most studies have shown that caseins and ß-lg are the main allergens in cow's milk (Cocco et *al.*, 2003). Fermentation by lactic acid bacteria could reduce some of the undesirable effects of these allergens. Various approaches have been used to reduce the antigenicity/allergenicity of these allergens. One possible approach would be to use certain physiological properties of milk hydrolysates fermented by lactic bacteria.

Despite the abundance of research into the probiotic benefits of lactic acid bacteria, questions remain as to their precise mechanisms of action and their impact on different biological models. Furthermore, the functional diversity of strains isolated from cow's milk and their specific influence on intestinal health, particularly in mouse models such as the BALB/c mouse, remain insufficiently explored. This book aims to fill these gaps by providing an in-depth analysis of the functional properties of lactic acid bacteria and assessing their role in modulating intestinal function. The central issue we propose to explore is as follows: to what extent can the functional characterisation of lactic acid bacteria derived from cow's milk contribute to a predictive model of the modulation of intestinal function in the BALB/c mouse?

1. Lactic acid bacteria

1.1 Properties and classification of lactic acid bacteria

Lactic acid bacteria (LAB) are one of the groups closest to humans. They are naturally associated with mucous membranes, particularly the intestine (Wood and Hozapfel, 1995; Wood and Warner, 2003).

The metabolic and physiological criteria for lactic acid bacteria enable them to be classified as gram-positive, catalase-negative, strictly fermentative bacteria that produce lactic acid as the main end product of sugar fermentation (Kandler, 1983).

LAB are also used in the industrial fermentation of dairy, meat and vegetable products. At this stage, the LAB group includes around 20 genera, of which *Aerococcus*, *Carnobacterium*, *Enterococcus*, *Lactobacillus*, *Lactococcus*, *Leuconostoc*, *Oenococccus*, *Pediococcus*, *Streptococcus*, *Tetragenococcus, Vagococcus* and Weissella are considered to be the genera most associated with foods (Axelsson, 2004).

Two main pathways for the fermentation of hexoses can be distinguished in lactic acid bacteria. These are glycolysis (Emden-Meyer pathway), which results in lactic acid almost exclusively as the final product (homofermentative) and the 6-phosphogluconate/phosphoketolase pathway, which results in the production of other products such as ethanol, acetic acid and CO_2 in addition to lactic acid (heterofermentative) (Schleifer and Ludwig, 1995).

1.2 Lactic acid bacteria in the production of fermented products

LAB are used in the production of a wide range of fermented dairy products such as cheese and yoghurt. They can contribute to the microbiological safety of the fermented product. The technological, nutritional and organoleptic quality results from the production of ethanol, acetic acid, flavours, exopolysaccharides, bacteriocins and several enzymes (Axelsson, 2004).

1.3 Fermentation process by lactic bacteria

The methods and knowledge associated with making fermented products have been passed down through several generations. By definition, fermentation is the process in which a substrate undergoes biochemical changes resulting from the metabolic activity of micro-organisms and their enzymes. During fermentation, energy (ATP) is derived from the partial oxidation of organic compounds, such as carbohydrates (Gotcheva et *al.*, 2000). The simple organic products formed from this oxidation process also serve as the final electron and hydrogen acceptor. ATP is produced by phosphorylation of the substrate.

At the end of the 1950s, Pasteur demonstrated that fermentation is a vital process specific to the growth of each microorganism, and each type of fermentation is defined by the end product formed (lactic acid, ethanol, acetic acid or butyric acid). During fermentation, pyruvate is metabolised into various compounds. Homolactic fermentation produces lactic acid, while heterolactic fermentation produces lactic acid and other acids and alcohols. Many food products owe their characteristics to fermentation and the activity of micro-organisms. Digestibility, nutritional value and organoleptic qualities are linked to shelf life and are enhanced by the fermentation process. Many ripened foods, such as cheeses and gherkins, are preserved because they have a longer shelf life than fresh products (Panesar et *al.*, 2011).

1.3.1 Metabolism of lactic acid bacteria in the fermentation of dairy products

1.3.1.1 Glycolysis

The first stage in the fermentation of lactic acid bacteria in cheese making is the fermentation of lactose into lactic acid. Acid production is very important in the development of organoleptic characteristics in certain types of cheese. A rapid reduction in pH during the initial stages of cheese preparation is essential for the coagulation of milk proteins. Acidity determines the pH value which influences the structure of the proteins as well as the structure and flavour of the product (Broadbent and Steele, 2005).

1.3.1.2 Citrate catabolism

Milk contains around 1.5g/l of citrate. A large amount is lost during the fermentation process and the rest is found in the soluble phase of milk. Nevertheless the low concentration of citrate is of great importance since it can be metabolised to a number of volatile compounds by mesophilic bacteria (Lactococci, Enterococci, leuconostoc. citrate positive) (McSweeney and sousa, 2000), Several studies have investigated the citrate metabolism of Enterococci which showed the high power to metabolise citrate. This property is used in the agri-food industry (Sarantinopoulos et *al.*, 2001; Rea and cogan, 2003).

1.3.1.3 Lipolysis

The formation of free fatty acids (FFA) by lipolysis of milk, and the conversion of free fatty acids to methyl ketone and tri-esters by lipases and esterases, have a direct effect on the process of developing the flavour and texture of the product. The enzymes involved in these reactions may come from the rennet, the milk itself or the lactic ferments. According to (Giraffa, 2003), esterases can be classified as enzymes that hydrolyse soluble substrates, whereas lipases hydrolyse insoluble substrates (Broadbent and Steele, 2005). The lactococcal lipases/esterases studied by several authors are intracellular (Fox et *al.*, 1993; Fox and Wallace, 1997). Lactobacillus strains such as *L. helveticus*, *L. delbruekii subsp bulgaricus* and *L.delbruekii subsp. lactis* also produce esterases, some of which have been studied (Khalid and Marth, 1990). Enterococci of food origin are the most lipolytic and esterolytic, in particular *E.faecalis*, followed by *E.Durans* and *Efaecium* (Tsakalidou et *al.*, 1993; Sarantinopoulos et *al.*, 2001). However, their lipolytic activity is conditioned by a number of factors and growth conditions. The esterolytic activity of enterococci is more effective than lipolytic activity (Giraffa, 2003).

1.3.1.4 Proteolysis

Proteolysis is considered to be one of the most important biochemical processes involved in the manufacture of many fermented dairy products. Proteolysis during cheese maturation results from the action of coagulating enzymes such as chymosin, and endogenous milk proteinases such as plasmin. Hydrolysis is initially initiated by the coagulant chymosin and endogenous proteinases. The second stage is bacterial proteolysis by proteinases and peptidases.

Low molecular weight peptides released from caseins directly affect flavour but some of them may also impart bitterness. The release of free amino acids can also directly affect flavour. For example, glutamate and aspartate residues enhance flavour and are taste stimulants. More commonly, released amino acids are precursors of many flavour compounds (Sarantinopoulos et *al.*, 2001).

1.3.1.5 Role of starter and non-starter lactic acid bacteria in fermentation

Older fermented products were prepared naturally in the presence of the microflora present in the original product. During such natural fermentation, the quality of the final product depends on the concentration of microorganisms and the fermentation conditions. The use of lactic ferments was developed to exploit the ferments to obtain successful products. This technique, known as re-inoculation, involves inoculating milk with small quantities of inoculum from a ferment, in order to optimise the spontaneous fermentation of different products (Williams et *al.*, 2000).

For many years, lactic acid bacteria have been used in the manufacture of fermented dairy products such as cheese and yoghurt. Today, lactic acid bacteria play an important role in the dairy industry because of their unique metabolism. These include the production of acid as a result of the metabolism of sugars, the proteolytic activities of various proteases and peptidases and also the production of antimicrobial compounds that can prevent or inhibit pathogens.

Starter lactic acid bacteria (BLS) are added at the beginning of the manufacturing process. Non-starter lactic acid bacteria (NSLAB) are those that have not been added as part of the starter culture, and use nucleic acids derived from the autolysis of BLS (Williams et *al.* 2000; Kieronczyk et *al.*, 2001; Diaz-Muniz et *al.*, 2006).

1.3.2 Proteolytic system of lactic bacteria

Lactic acid bacteria are unable to synthesise the amino acids they need to grow (fig.1). Milk proteins are broken down into peptides and amino acids during the fermentation process, which are used as essential nutrient sources for growth (Juillard et *al.*, 1995; Hafeez et *al.*, 2014).

The proteolytic system of lactic acid bacteria is mainly composed of **i)** one or more cell wall proteases also known as cell envelope proteinases capable of hydrolysing milk proteins into peptides between 4-30 residues (Laan & Konings, 1989); **ii)** the proteolytic system includes oligopeptides, two permeases acting as pores and two ATPases which provide energy and **iii)** intracellular peptidases necessary for peptide degradation. Cell envelope proteinases are initially responsible for hydrolysis and are the main players in the release of bioactive peptides (Savijok et *al.*, 2006).

Five types of proteinases have been characterised in Iactobacilli (Holck and Naes, 1992; Gilbert et *al.*, 1996; Pederson et *al.*, 1999; Siezen, 1999; Fernandez-Espla et *al.*, 2000; Pastar et *al*, 2003) including PrtP for *Iactococcus lactis* and *lactobacillus paracasei*, PrtS for *streptococcus thermophillus*, Prt H for *lactobacillus helveticus*, Prt B for *L. delbruekii subsp bulgaricus* and PrtR for *L. rhamnosus*. Proteinases do not contain the same number of domains. For example, domain B is absent on Prt S, domain I on Prt R and domain AN on Prt H and Prt B (Exterkate et *al.*, 1993). They are all classified as subtilisin serine proteinases because of the homology between their catalytic domains and the subtilisin of bacillus subtilis. *L.lactis* has a number of different proteinases, distinguished by their modes of action (Exterkate et *al.*, 1993). Type PI hydrolyses β-casein, more than 100 oligopeptides ranging in size from 4 to 30 amino acid residues. Type PIII capable of hydrolysing Γα-S1 and β-casein. Proteinases have intermediate characteristics between PI and PIII.

The two types (I and III) of proteinases share 98% similarity with five substitutions in the substrate binding region of these two proteases (Exterkate et *al.*, 1993). Several parameters such as enzyme/substrate ratio, medium composition, heat treatment, temperature, pH, carbon/nitrogen ratio influence the expression and ability of proteinases to release bioactive peptides from milk proteins (Hafeez et *al.*, 2014). Different hydrolysis profiles have been obtained in strains with the same prtS-type proteinases under different conditions (Miclo et *al.*, 2012).

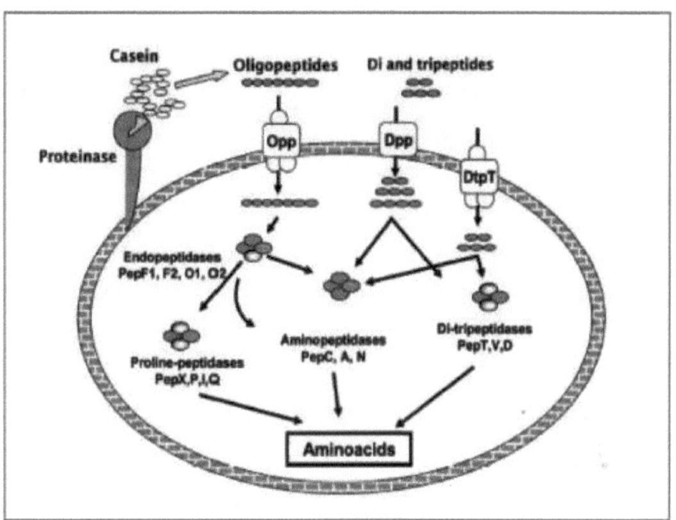

Fig.1: Schematic presentation of the proteolytic system of lactic acid bacteria (Kunji et *al.*, 1996)

2. Milk proteins

2.1 Composition

Milk contains 30-35g of protein per litre, representing 95% of total nitrogen. Around 80% of the protein is in the form of a spheroidal supramolecular structure containing a mineral component known as casein micelles (Table 1). These are easily separated by ultracentrifugation in their native form, by isoelectric precipitation at pH 4.6 of the casein in a denatured form, or by goagulation of the milk by chymosin (rennet coagulating enzyme) in the form of a phosphocalcium complex (Barnig et al., 2005).

The non-sedimentable (or filterable) fraction known as whey contains around 18% of total proteins. These are known as soluble proteins or whey proteins. They contain two major globular proteins, ß-lactoglobulin and a-lactalbumin, and other proteins in smaller quantities (immunoglobulin, seralbumin, etc.). The four caseins (α_s 1, as2, β and κ) and the two major whey proteins (β- lactoglobulin and a- lactalbumin) are synthesised by the mammary gland, the others being mostly of blood origin; their primary sequence is known and so is their spatial structure (Sicherer et al., 1999).

Caseins are molecules with a poorly ordered, highly unfolded structure, with polar side chains (mainly phosphoserylated and carboxylic) grouped on the N-terminal side (as1 and β) or apolar on the C-terminal side ("si and β), this distribution being reversed for κ casein (polar on the C-terminal side). It should be noted, however, that as2 casein exists partly in dimeric form (intermolecular disulphide bridges) and that κ casein may be present in the form of monomers or polymers (degree of polymerisation may extend beyond 10 units) (Lorient and Cayot, 1991).

While the micelle is very stable at high temperatures above 100°C and to mechanical treatment, it is very sensitive to changes in the environment: acidification to pH 4.6 destabilises the mineral phase and causes precipitation of the demineralised casein. Low temperatures also cause partial destabilisation of the micelle, which loses part of the β-casein. Soluble proteins have a compact globular structure, stabilised by disulphide bridges.

Table 1: Structural characteristics of the main milk proteins (Lorient and Cayot, 1991)

Fraction	Concentration g/l (% protein)	Prosthetic part P(phosphorus) G(carbohydrate)	Disulphide bridge (or SH group)	Molecular weight	structure
Caseins	24-28 (82)				micellar
Otsl	12-15 (39-46)	8P	0	23612	bipolar
αs2	3-4(8-11)	10-13P	2cys SH	25228	unfolded
ß	9-11(25-35)	P	0	23980	bipolar
K	3-4(8-15)	1P-1G	2cys SH	19000	bipolar
γ	1-2(3-7)				β-casein fragments
Soluble proteins β-lactoglobulin α-Lactoglobulin serum albumin immunoglobulin	5-7 (18) 2-4 (7-12) 1-1,5(2-5) 0,1-0,4(0,7-1,3)	Calcium	2SS-1SH 4S-S 17S-S-1 SH	18362 14174 69000	Compact Globular Globular oligomers Globular Oligomers

protein-peptones	0,6-1(1,9-3,3) 2-5 (0,6-1,8)			15000 à 1.000.000	Globular or spreading

denatured by heat, causing loss of solubility or gelation, depending on the pH and the concentration of protein and mineral salts. Because of their compact structure, and unlike micelles, they are highly resistant to certain proteolyses in their native state. Their physico-chemical properties are easily modified by denaturation (Considine et *al.*, 2007; Anema, 2008).

2.2 Functional properties of milk proteins

Milk proteins play an essential role in our daily diet, as they are consumed in large quantities in a wide variety of forms, such as drinking milk, dairy products (cheese, yoghurt, dairy desserts) and numerous food preparations (preserves, ready meals, sauces, soups, pastries, confectionery). Their balanced composition in terms of essential amino acid residues and their good digestibility are their main assets in the eyes of nutritionists (Korhonen et *al.*, 2006). Added to this is the fact that these proteins, with their varied molecular structures, give the foods in which they are incorporated excellent sensory acceptability thanks to a large number of physico-chemical and techno-functional properties. They constitute polyfunctional ingredients with excellent added value that new fractionation techniques are capable of supplying to the food industry in forms adapted to different uses (Sicherer et *al.*, 1999).

2.2.1 Intestinal absorption of proteins

The intestinal epithelium, which covers the digestive mucosa, is an important interface between the intestinal lumen and the mucosal immune system and constitutes a communication site where the most complex interactions between genes and the immune system occur (fig.2). It allows nutrients to be absorbed while forming an effective barrier to prevent the massive entry of incompletely hydrolysed food antigens (Brandtzaeg, 2011). The intestinal barrier essentially consists of an epithelial monolayer comprising absorptive enterocytes, mucus cells, endocrine cells, Paneth cells and M cells located above Peyer's patches and isolated lymphoid follicles specialised in the controlled transmission of micro-organisms to the immune cells of the chorion (Spahn and Kucharzik 2004) (fig.3).

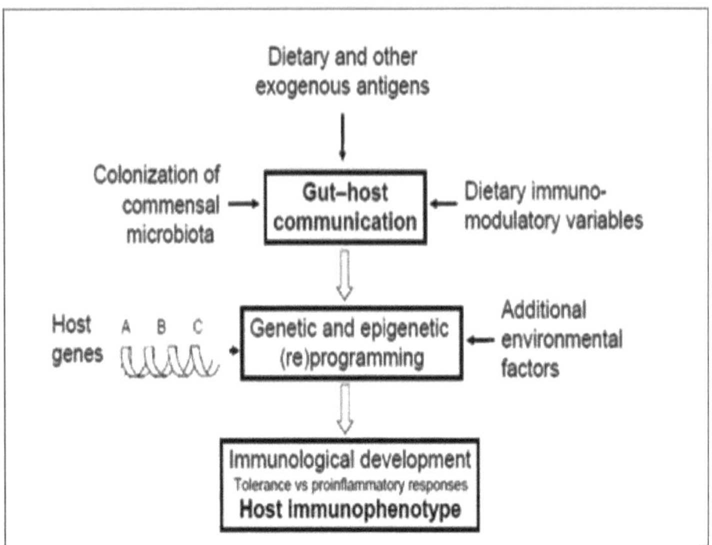

Fig.2 Influence of external antigens on the gastrointestinal tract (Brandtzaeg, 2011)

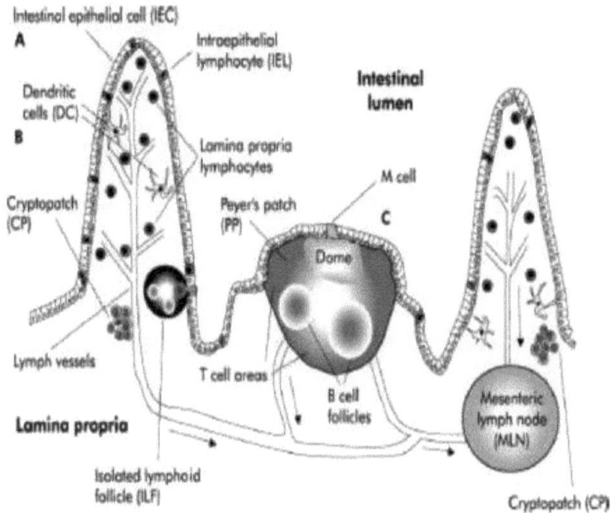

Fig.3 Schematic representation of the lymphoid elements of the GALT

CP: crypto-plates; IEL: intraepithelial lymphocytes; PP: Peyer's patches; MLN: mesenteric lymph nodes; ILF: lymphoid follicles. Antigens from the intestinal lumen are taken up by 4 main routes: directly by enterocytes: IEC (A), by dendritic cells: DC (B), by M cells (C) or by a paracellular route not shown in the diagram (D). (Spahn and Kucharzik 2004)

At the level of the intestinal barrier, the passage of intact food proteins is a necessary step for immunological recognition. The digestive tract and the gastrointestinal tract-associated lymphoid system (GALT) play a major role in sensitisation to food allergens (Adel-Patient et *al.*, 2008). Under physiological conditions, few food proteins are absorbed intact and arrive at the GALT in an immunologically active form. Some food allergens are resistant to digestion, reach the site of induction of an immune response and cause sensitivity in the allergic subject (Adel-Patient et *al.*, 2008; Zellal, 2009, El mecherfi et *al.,* 2015). This is the case for bovine milk ß-lactoglobulin, which is resistant to pepsin (Moreno, 2007). Peptides of different molecular weights can retain the allergenicity of the native protein even after degradation during the digestion process (Zellal, 2009). One study showed that the allergenicity of ß-Lg increased after proteolysis following the revelation of hidden epitopes within the tertiary structure of the protein (Selo et *al.*, 1999).

2.2.2 The different pathways involved in intestinal permeability

The limiting factor in the diffusion of antigenic or toxic molecules from the intestinal lumen to the underlying chorion is the monolayer of epithelial cells. Two absorption pathways may be involved in their intestinal transfer: the paracellular pathway and the transcellular pathway (fig. 4) (Keita and Soderholm, 2010).

In physiological conditions, paracellular diffusion only concerns small molecules with a molecular mass of less than 600 daltons (inert permeability markers such as lactulose and mannitol in vivo or 51Cr-EDTA in vitro). Intercellular tight junctions are the major structure limiting paracellular permeability (Powell, 1981; Desjeux et *al.,* 1984; Marcon-Genty et *al.,* 1989; Adson, 1994; Ménard et *al.*, 2012).

There is an increase in absorption capacity via the paracellular route in an inflammatory environment. This involves a complex regulation linked to changes in the structure of the tight junctions (remodelling by phosphorylation of the constituent proteins, occludin, claudins, JAM-A, ZO1, 2, 3) observed in inflammatory conditions. Ionic movements are generally responsible for the movement of water and soluble molecules and the phenomenon of solvent drag (Ménard, 2010).

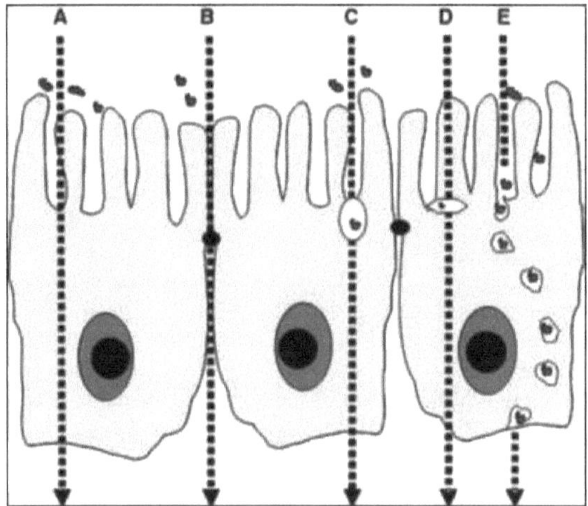

Fig. 4 Transepithelial transport pathways.

A. The transcellular pathway; B. The paracellular route; C. The transcellular pathway via aquaporins (passive exchange
aqueous pores); D. Active transport (nutrients); E:
Endocytosis-transcytosis-exocytosis (Keita and Soderholm, 2010).

There is also a transcellular transport pathway for luminal antigens, involving an active transcytosis mechanism (internalisation of the apical membrane and luminal contents, formation of endosomes, migration of endosomes and exocytosis of degradation products to the basal membrane) (Ménard, 2010).

However, the quantity of antigen crossing the epithelial barrier is very small (2 $\mu g/h/cm^2$ of mucosa) and represents only 1/1000 of the luminal concentration (if the luminal concentration of an antigen is 1 mg/ml, a concentration of 1 $\mu g/ml$ will be observed in the chorion). This phenomenon of transcytosis of dietary proteins is therefore of no nutritional interest, but it is necessary to inform the mucosal immune system (Heyman et *al.*, 1984). Proteins and peptides reaching the intestinal lamina propria can be captured by local antigen-presenting cells, as shown by the study by Chirdo et *al* (2005) indicating that after gavage of mice with ovalbumin, this antigen is rapidly associated with intestinal dendritic cells (Ménard, 2010).

2.2.3 Role of epithelial exosomes in informing the intestinal immune system

Antigens such as food proteins and micro-organisms present in the mother's diet can be recognised by the lymphocyte system of the digestive system. More specifically, naive lymphocytes are grouped together in particular free structures called Peyer's patches. At this level, the barrier between the digestive contents and these lymphocytes is partly made up of M cells, which have both an endocytosis function and a lack of proteolytic activity, as they have virtually no lysosomal system. The lymphocytes are therefore directly stimulated by luminal antigens. The result is clonal multiplication and migration towards the various mucosal structures of the organisms, including the mammary glands of breastfeeding women (El mecherfi, 2012).

Since a significant fraction of food antigens are transmitted across the intestinal barrier in the form of immunogenic peptides, this suggests incomplete degradation and 'protection' of the peptides during transepithelial transport. The notion that antigen-presenting dendritic cells can prime proteins into peptides and release vesicles (exosomes) bearing on their surface these peptides associated with major histocompatibility complex (MHC) class II molecules led to the investigation of the possibility of exosome production by intestinal epithelial cells (Fig.5) Ménard, 2010). This hypothesis was supported by the demonstration of the secretion of vesicular structures (80 nm in diameter) by intestinal epithelial cell lines whose structure and molecular composition are close to those of professional antigen-presenting cells (Raposo, 1996).

Exosomes are formed by internalisation of the outer membrane of this compartment, which explains the presence of MHCII/peptide complexes on their surface.

MIIC compartments can either fuse with the lysosomal system or fuse with the plasma membrane and release their exosome contents into the extracellular environment. This phenomenon has been demonstrated in intestinal epithelial cells (Van Nieg et *al.*, 2001) and could be important for the transfer of luminal antigens in highly immunogenic form (Heyman, 2010). Indeed, in vitro, peptides derived from antigen priming by the epithelial cell and released in the form of exosomes can interact very effectively with cultured dendritic cells and stimulate specific T clones at concentrations 100 times lower than those required for activation by free peptides (Mallegol, 2007).

Transcytosis of food antigens is essentially carried out by non-specific 'fluid phase' endocytosis. However, in certain circumstances, luminal antigens can reach the intestinal mucosa (Heyman, 2010). In the form of immunocomplexes (fig.6), thanks to the expression of immunoglobulin (Ig) receptors expressed in normal or pathological conditions on the apical surface of enterocytes. These may be IgE immunocomplexes in the case of food allergies or IgA/gliadin complexes as recently described in celiac disease.

3. Cow's milk protein allergy

In children, cow's milk is one of the three most frequent food allergens, along with eggs and peanuts. It is responsible for 16% of food allergies (Bidat, 2006). In infants, cow's milk protein (CMSP) is the first and only food antigen introduced into the diet until diversification. This is why cow's milk protein allergy (CMPA) is a condition that occurs early, mainly in the first few years of life (Fiocchi et *al.*, 2010), with a prevalence of 2 to 3% (Host et *al.*, 2002). Among these antigens, ß-lactoglobulin (ß-Lg), a-lactalbumin (α-La), and caseins are the main allergens in cow's milk. Bovine serum albumin (BSA) and even lactoferrin present in minute quantities are also potential allergens (Sharma et *al.,* 2001; Fritsche, 2003; Bu et *al.,*2013).

In Algeria, the incidence is 1.8% (Ibsaine et *al.*, 2013). Recent studies have shown that changes in the natural history of allergy have appeared over time and that the onset of the disease is increasingly slow, with persistence into adolescence and adulthood (Skripak et *al.*, 2007).

3.1 Mechanism of cow's milk protein allergy

Three mechanisms are involved in the immune-mediated reaction (Morali, 2004): IgE-dependent allergy, non-IgE-dependent allergy and mixed IgE/non-IgE-dependent allergy (Fox and Thomson, 2007).

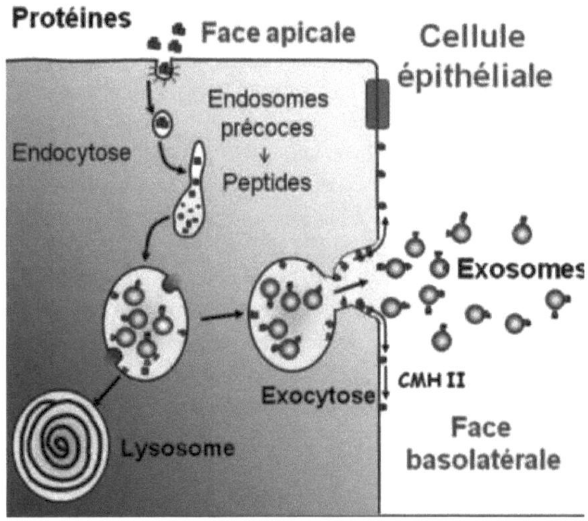

Fig.5 Exosomes are antigen-presenting vesicles carrying major incompatibility complex II/peptides on their surface (Ménard, 2010).

3.2 IgE-mediated intestinal transport of allergens in allergy

The low affinity receptor for IgE (FcεRI, CD23) is capable of transporting IgE immune complexes across the enterocyte in intestinal allergy (fig.7). CD23 is a receptor mainly expressed on haematopoietic cells but its expression has also been observed on the apical and basal surfaces of enterocytes in patients with intestinal diseases, whether IgE-dependent or not (cow's milk protein allergy, autoimmune enteropathy, Crohn's disease, haemorrhagic rectocolitis) (Kaiserlian, 1993; Lachaux, 1996).

High expression levels of IL-4, a Th2 cytokine involved in allergic diseases, are responsible for the overexpression of CD23. Although IgE is not considered to be an immunoglobulin secreted in the intestinal lumen, it is found in intestinal washings during parasitic infection (Negrao-Correa, 1996) or in food allergy (Belut, 1980).

The role of epithelial CD23 and IgE immuncomplexes in the entry of food allergens into the intestinal mucosa has been demonstrated in mouse models of allergy.

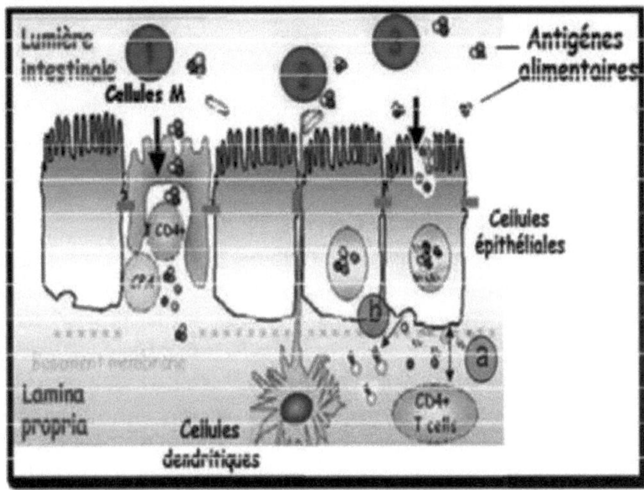

Fig.6 Management of food antigens present in the intestinal lumen and their presentation to the immune system associated with the intestinal mucosa

(Mallegol et *al.*, 2005).

The transepithelial passage of antigens present in the intestinal lumen and their presentation to the GALT immune system involves three different pathways: (1) capture by M cells covering the peyer's patches (2) direct sampling by dendritic cells in the lamina propria, via dendrites extending into the intestinal lumen, and (3) endocytosis of antigens by epithelial cells and subsequent presentation to underlying T cells (a) or release of exosomes carrying antigenic peptides associated with major histocompatibility complex molecules (b).

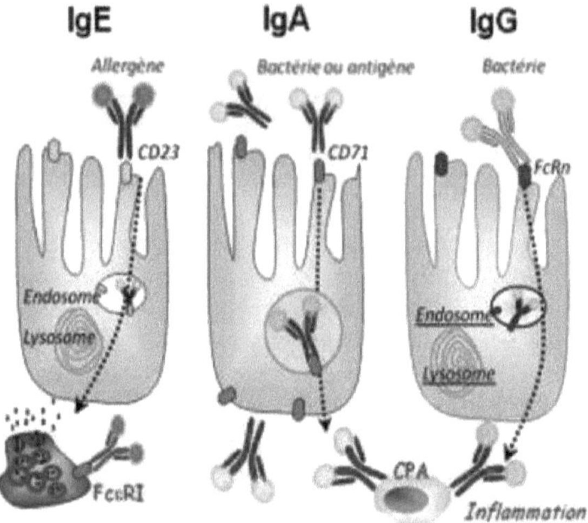

Fig. 7 IgE, IgG and especially IgA immunoglobulins are present in the intestinal lumen (Heyman, 2010).

Sensitisation of rats to a test protein, horseradish peroxidase (HRP), led to this protein being absorbed into the enterocytes and transported more rapidly than that observed in control animals (Heyman et *al.*, 1984; Yang, 2000).

This rapid transfer involves the CD23 receptor and the presence of IgE immunocomplexes (Berin, 1997). Thus, allergic sensitisation by increasing the expression of the CD23 receptor allows an IgE-complexed allergen to cross the epithelial barrier in intact form, avoiding lysosomal degradation. This mechanism enables IgE-allergen immunocomplexes delivered to the chorion to rapidly induce mast cell degranulation, leading to an inflammatory allergic response.

After repeated exposure to food allergens, atopic subjects develop a powerful TH2 response and produce type E immunoglobulins. IgE subsequently binds to mast cells and basophils via their FcεRI receptors (Galli, 2000; Kawakami, 2002; Galli, 2005). Re-exposure to the same antigens leads to activation and degranulation of mast cells and basophils, which, after bridging the allergen with surface IgE, secrete inflammatory mediators: cytokines and chemokines, resulting in a type I immediate hypersensitivity inflammatory response (Cole, 2001; Wood, 2003; Bischoff, 2005). IgE antibodies are the only antibodies that regulate the expression of the FcεRI receptor on mast cells and basophils, which leads to the amplification of the type I immediate allergic reaction (Lantz, 1997).

3.2.1 IgE-dependent hypersensitivity (immediate reaction or type I)

In the IgE-dependent reaction, allergic symptoms are most often immediate. They are linked to degranulation of mast cells, with release of mediators, after contact with a specific milk antigen (Rance, 2007).

This type I allergy is based on the formation of immunoglobulin E (IgE) against allergenic milk proteins (fig.8). In this type of allergic reaction, cutaneous, intestinal or respiratory symptoms appear within two hours of milk ingestion (de Boissieu and Dupont, 2006; Tsuge et *al.*, 2006).

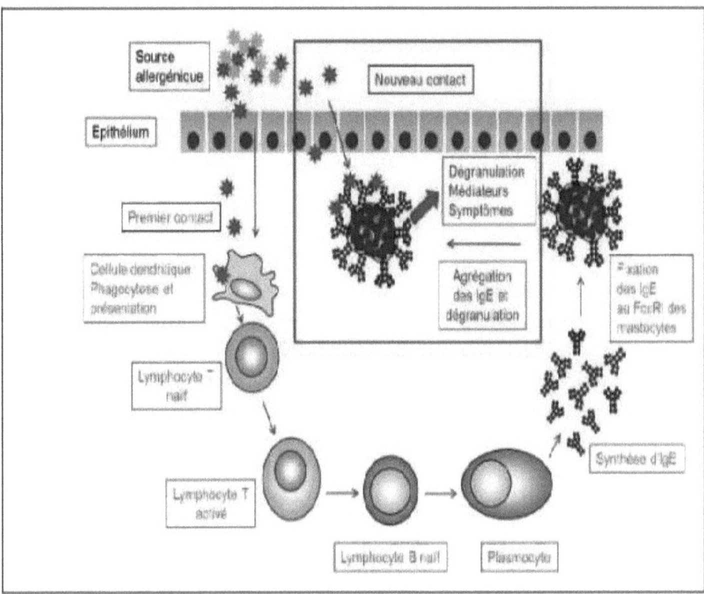

Fig.8 IgE-dependent hypersensitivity (immediate reaction or type I) (Larché et *al.*, 2006)

These symptoms are triggered when an allergen binds two IgE antibodies with high affinity to FCεRI receptors on the surface of a mast cell or basophilic cell, inducing degranulation of the cells and the release of biological mediators such as histamine, as well as mediators responsible for inflammatory reactions (Merja et *al.*, 2007). Th2 cells, which produce IL-4, IL-5, IL-10 and IL-13, control type I allergic reactions by activating B lymphocytes and regulating IgE secretion (Botturi and Magnan, 2006).

3.2.2 Semi-delayed hypersensitivity (involvement of circulating immune complexes or CIC) (Type III)

During semi-delayed hypersensitivity, specific IgE or IgG immune complexes may form in the serum with antigens (cow's milk proteins) at either local or systemic sites, leading to phagocytosis or complement-mediated injury (Lydyard et *al.*, 2002). This reaction may appear after 8 to 12 hours of milk ingestion (Morali, 2004).

3.2.3 Non-IgE-dependent hypersensitivity (delayed reaction or type IV)

This hypersensitivity reaction, the only type transferable by T cells instead of antibodies, appears at least 24 hours after contact with a provocative antigen (milk proteins). It is manifested by cutaneous and intestinal symptoms (Tsuge et *al.*, 2006). Responses to this hypersensitivity often lead to the production of granulomas a few weeks later. This hypersensitivity also plays a role in a number of pathologies where there is persistence of the antigen that the immune system cannot suppress, thus producing chronic inflammation (Lydyard et *al.*, 2002).

Th1 cells are involved in this delayed hypersensitivity reaction and induce inflammation, predominantly secreting IL-2, IFN-γ and TGF-β. Tc cells induce cytotoxic responses (Botturi and Magnan, 2006). In addition to T cells, the main players in this type of sensitivity are dendritic cells, macrophages and cytokines (Lydyard et *al.*, 2002).

3.3 The role of lactic acid bacteria in reducing the antigenicity and allergenicity of cow's milk proteins

During fermentation, lactic acid bacteria have the ability to hydrolyse protein epitopes, thereby reducing the allergenicity of the resulting products. Bacterial polysaccharides, or hydrocolloids, commonly used as texturisers in the food industry also affect protein digestibility (Kleber et *al.*, 2006). By studying these hypotheses, researchers have shown that cells of the lactic acid bacterium *L. acidophilus* are capable of hydrolysing a milk protein, β-lactoglobulin, at a rate of 52%, whereas the degradation of this protein by pepsin, an enzyme in gastric juice, is low (8%). This rate rises to 55% when pepsin hydrolysis is preceded by pre-hydrolysis by *L. acidophilus*, and even to 58% in the presence of pectin or exopolysaccharides. As the main epitopes of β-Iactoglobulin have been hydrolysed, the allergenicity of this protein could be reduced. These results suggest that *L. acidophilus* CRL 656 could be used as a culture additive during the fermentation of milk or whey in the manufacture of fermented dairy products with hypoallergenic properties (Pescuma et *al.*, 2011). It is now accepted that food allergies occur when oral tolerance is not induced or maintained over time. Therefore, by stimulating the induction of oral tolerance, we can prevent the development of allergies (Besler et *al.*, 2001; Ruttarattanamongkol, 2012).

4. Lactic acid bacteria in human health

4.1 Fermented milk hydrolysates and the health benefits of peptides

Over the last two decades, interest in the use of dairy protein hydrolysates containing bioactive peptides has been developed for the prevention of certain chronic diseases (Hernández-Ledesma et *al.*, 2014).

Bioactive peptides derived from milk during fermentation usually consist of dipeptides or oligopeptides, which in some cases can exceed 20 amino acid residues (Wang & De Mejia, 2005; Erdmann et *al.*, 2008; El fahri et *al* .,2014). In relation to size and amino acid content, these peptides can exert different physiological activities such as antihypertensive activity, antioxidant activity and immunomodulatory activity (fig.9). Recently, interest has focused on high-potency dietary peptides that can reduce oxidative stress (Xiong, 2010; Zhao, Wu, & Li, 2010; Moslehishad et *al.*, 2013).

4.2 Production of bioactive peptides

Basically, biologically active peptides can be produced from milk proteins following: **(a)** digestive enzymatic hydrolysis, **(b)** fermentation (proteolysis) (table 2), **(c)** by enzymes from micro-organisms or plants.

The combination of **(a, b, c)** generates functional peptides (Korhonen & Pihlanto, 2003b).

4.3 Production of bioactive peptides by lactic acid bacteria

Due to the inability of lactic acid bacteria to synthesise all the amino acids required for growth, lactic acid bacteria break down milk proteins into peptides and amino acids during the fermentation process, using them as a source of growth (Juillard et *al.*, 1995). These peptides can have beneficial effects on health. The proteolytic system involves **i)** one or more proteases called cell envelope proteases (CEP) (Laan & Konings, 1989) capable of hydrolysing milk proteins into peptides, **ii)** a transport system through two permeases with energy release, **iii)** intra-cellular peptidases for the degradation of peptides into amino acids (Savijoki et *al.*, 2006). PECs are initially responsible for the hydrolysis of milk proteins and the release of bioactive peptides (Siezen, 1999; Fernandez-Espla et *al.*, 2000; Hafeez et *al.*, 2014). The *L. helveticus* strain has been widely studied in the production of fermented dairy products such as cheeses. This strain has a high proteolytic capacity and is capable of releasing angiotensin-inhibiting peptides (Nakamura et *al.*, 1995; Pihlanto-Leppala et *al.*, 1998; Sipolae *al.*, 2002).

4.4 Production of bioactive peptides by digestive enzymes

One of the most common techniques for producing bioactive peptides from milk proteins is enzymatic hydrolysis. Many bioactive peptides have been obtained after hydrolysis by digestive enzymes, such as trypsin and pepsin (Meisel & FitzGerald, 2003; Yamamoto et *al.*, 2003; FitzGerald et *al.*, 2004; Gobbetti et *al.*, 2004; Vermeirssen, et *al.*, 2004). These studies have reported the potency of peptides with antihypertensive, immunomodulatory, antioxidant and antimicrobial action.

4.4.1 Antihypertensive activity

Angiotensin-converting enzyme (ACE) has been associated with the renin angiotensin system, which regulates peripheral blood pressure (Korhonen & Pihlanto, 2006; De Leo et *al.*, 2009; Moslehishad et *al.*, 2013). Inhibition of this enzyme may have an antihypertensive effect. A large number of peptide groups with antihypertensive activity have been isolated after enzymatic hydrolysis of milk proteins. At present this is the most studied group of bioactive peptides (Hafeez et *al.*, 2014). The antihypertensive effect of bioactive peptides consists of inhibiting the conversion of angiotensin I to angiotensin II (Vermeirssen, et *al.*, 2004). These peptides have already been isolated from a variety of fermented dairy products including cheeses (Hartmann & Meisel, 2007), yoghurts (Donkor, 2007) and fermented bovine milk (Qian et *al.*, 2011) (table 3).

Peptides with antithrombotic activity

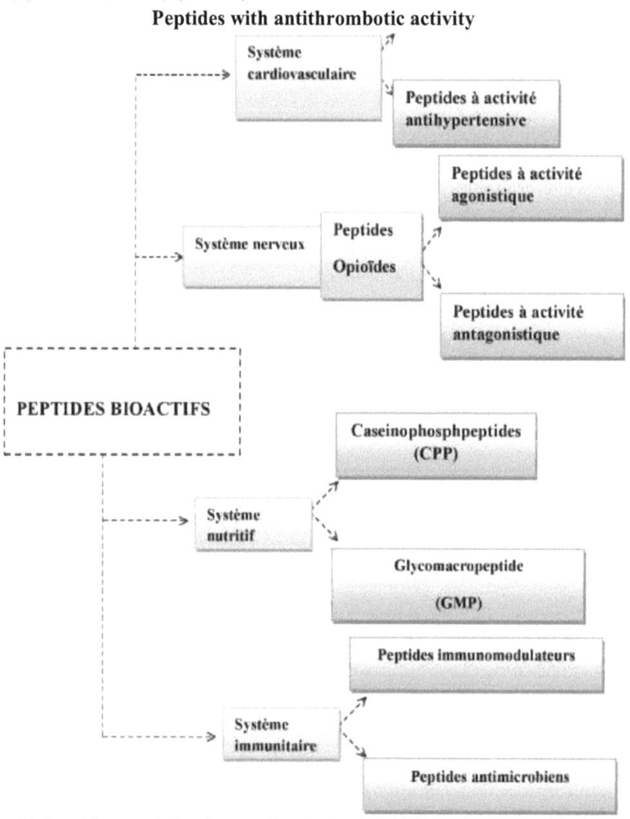

Fig.9 Roles of bioactive peptides in caseins (Silva and Malcata, 2005; Hafeez et *al.*, 2014)

Table 2: Example of the bioactivities of some fermented milks (Hafeez et *al.*, 2014)

Type of product	Bioactivities	References
Yogurt (traditional and commercial	Antioxidant activity	Aloglu and Oner (2011)
Yogurt(probiotic)	Antioxidant , antidiabetic activities	Ejtahed et *al* (2012)
Fermented milk (L. delbrueckii subsp. bulgaricus LA2)	ACE-inhibition	Regazzo et *al* (2010)
Fermented milk (L. delbrueckii subsp. bulgaricus LA2)	Immunomodulation	Regazzo et *al*, (2010)
Fermented milk (E. faecalis TH563)	ACE-inhibition	Regazzo et *al*, (2010)
Kefir (milk & soyamilk)	Hypocholesterolemic	Liu et *al.*, 2007

Table 3: Some antihypertensive peptides present on the primary structures of caseins (Silva and Malcata, 2005).

Casein type Amino-acid	Peptide sequence	segment	References
α-s1 (bovine)	23-24	FF	Maruyama and Susuki (1982); Maruyama et *al.* (1987a);
	102-109	KKYKVPQ	Gomez-Ruiz et *al* (2002)
a-s2 (bovine)	189-197	AMKPWIQPK	Maeno, Yamamoto, and Takano (1996)
	174-179	FALPQY	Tauzin, Miclo, and Gaillard (2002)
β (bovine)	74-76	PPI	Nakamura et *al* (1995)

4.4.2 Immunomodulating activity

Milk protein hydrolysates and peptides derived from caseins and whey can improve immune cell function and regulate cytokine synthesis (Sutas et *al.*, 1996; Gill et *al.*, 2000; Matar et *al.*, 2003; Meisel & FitzGerald, 2003). The protective effect of a casein-derived immunopeptide on resistance to microbial infection by *Klebsiella pneumoniae* has been demonstrated (Migliore-Samour et *al.*, 1989). It has also been suggested that immunomodulatory peptides can relieve allergic reactions in atopic individuals and boost the immunity of the intestinal mucosa (Korhonen & Pihlanto, 2003). In this context, immunomodulatory peptides can regulate the development of the immune system in newborns. It has recently been shown that commercially available whey proteins contain immunomodulatory peptides that can be released after enzymatic digestion (Mercier et *al.*, 2004). In addition, several studies have shown that immunopeptides formed during milk fermentation have anti-tumour effects (Matar et al., 2003)(table 4).

Table 4: Some sequences of immunomodulatory peptides present on the primary structures of caseins (Silva and Malcata, 2005).

Casein type	Peptide sequence	Amino-acid segment	References
α-s1 bovine)	1-23	RPKHPIKHQGLPQEVLNENLLRF	Minkiewicz et al (2000)
α-s2 (bovine)	1-32	KNTMEHVSSSEESIISQETYKQEKNMAI NPSK	Hata et al (1999)
β (bovine)	1-28	RELEELNVPGEIVESLSSSEESITRINK	Hata et al (1999)

4.4.3 Antioxidant activity

Recent studies have shown the antioxidant effect of peptides following hydrolysis of caseins after enzymatic digestion or fermentation by proteolytic bacteria (Korhonen & Pihlanto, 2003; Hafeez et al., 2014). Bacterial fermentation has been shown to be one of the most economical and practical methods for producing fermented dairy products rich in bioactive peptides (Hayes et al., 2007; Moslehishad et al., 2013). Recently, peptides with antioxidant properties have been obtained by enzymatic or bacterial hydrolysis of soya milk (Apostolidis et al., 2006; Elias et al., 2008; Zhao et al., 2010;Shori et al., 2013). Identified peptides released from the as-casein fraction have anti-free radical properties capable of inhibiting enzymatic and no-enzymatic lipid peroxidation (Suetsuna et al., 2000; Rival et al., 2001).

Many peptides derived from milk proteins can produce several physiological effects. For example, peptides in the 60-70 sequence of β-casein have an immunomudulatory and opioid effect and inhibit the enzyme that converts engiotencin (ACE) (Migliore-Samour & Jolle's, 1988; Meisel, 1997). Lipid oxidation can generate free radicals and plays a significant role in cardiovascular disease (Maxwell & Lip, 1997).

In recent years, new classes of peptides have been identified as antioxidants. Lipid oxidation is one of the main causes of food spoilage and reduced shelf life in the food industry (Pihlanto, 2006).

CHAPTER 5

Materials and methods

1. Properties of lactic acid bacteria isolated from cow's milk

1.1 Sampling and isolation of lactic acid bacteria

Five samples of cow's milk were collected in western Algeria under sterile conditions. For each sample of Rayeb milk traditionally fermented at room temperature, 10 ml were sterile pipetted into 90 ml of saline solution (0.9% NaCl (W/V)) and mixed. Decimal dilutions were made in the same solution up to 10^{-8}. One millilitre of each dilution was inoculated onto solid SRM. The plates were incubated at 37°C for 48 hours. Petri dishes containing between 30 and 300 colonies were selected for purification.

1.2 Biochemical identification of lactic acid bacteria

After purification on MRS (De Man et *al.*, 1960) and M17 (Terzaghi et Sandine, 1975) media, the isolates were examined under a microscope and tested for Gram staining and catalase activity. Purified isolates were stored at -20°C in sterile reconstituted skimmed milk (12.5%, W/V) with 30% (W/V) glycerol.

A strain of *Bifidobacterium longum* (*Blg*) was also used in this study, isolated from infant stools and belonging to the Nutrition Physiology Laboratory. It was regenerated twice in cysteine SRM and incubated anaerobically at 37°C.

1.2.1 Gram staining

Gram staining is the most widely used stain in bacteriology, enabling the properties of the bacterial wall to be highlighted and used to distinguish and classify bacteria. It is a differential stain that allows bacteria to be classified into two groups: Gram-positive bacteria fix the crystal violet and appear purple, and Gram-negative bacteria do not fix the crystal violet and appear pink (Larpent et larpend, 1990).

1.2.2 Catalase test

Catalase is an enzyme that catalyses the release of water and oxygen molecules from hydrogen peroxide, a molecule that is toxic to bacteria. The catalase test provides an initial orientation in the classification of a pure bacterial strain. LAB do not contain catalase. A drop of 3% (V/V) hydrogen peroxide (H_2O_2) is added to a colony placed on a glass slide. The reaction was positive when gas was observed.

1.2.3 Biochemical identification by api gallery

To determine the species, the fermentative profile of carbohydrates was studied on an API 50 CH and API 20 STRP (Biomerieux) gallery.

The 50 CH gallery is made up of 50 microtubes, and the STRP gallery is made up of 20 microtubes. Both allow the study of the fermentation of substrates belonging to the carbohydrate family and derivatives (heterosides, polyalcohols, uronic acids). The inoculum is prepared in MRS medium (without meat extract), then distributed using a sterile micropipette in the 50 and 20 tubes of the two galleries. During the incubation period, fermentation is reflected by a change in colour in the tube, due to anaerobic acid production revealed by the pH indicator (bromocresol purple).

The first microtube is used as a control. The results are read after 24 h and 48 h of incubation. The results are discussed according to Carr et *al* (2002).

These strains were identified biochemically. One has been identified by 16S DNA *PCR*. The others are currently being identified.

1.3 Identification by 16S DNA PCR

The Enterococcus strain was identified using the specific Enterococcus primer gene Ent 1(5'-TACTGACAAACCATTCATGAT-3') and Ent 2(5'AACTTCGTCACCAACGCCAAC- 3') (Ke et al., 1999).

The universal primers fD1 (5'- AGAGTTTGATCCTGGCTCAG-3') and rD1 (5'-TAAGGAGGTGATCCAGGC-3') were used (Weisburg et al., 1991) in a DNA thermal cycler (Techno, Barloworld Scientific, Cambridge, UK). PCR buffer (20 mM Tris-HCl 50 mM KCl, pH 8.4), 1.5 mM MgCl2, 0.2 mM of each dNTP, 1 U Taq DNA polymerase (Qiagen), 1 mM of each primer and 40 ng DNA were used in a final volume of 50 μl. PCR amplification was carried out under the following conditions: denaturation at 94°C for 5 min, 35 cycles of denaturation at 94°C for 1 min, at 56°C for 1 min, 15 DNA extensions at 72°C for 1 min. A final extension was added at 72°C for 5 min. Amplified DNA was analysed on a 1% (w/v) agarose gel with ethidium bromide (0.5 mg/ml) at 0.5 x TAE (40 mM Tris-acetate, 1 mM EDTA) buffer pH 8.2-8.4, for 30 min at 100 V and made visible by UV trans-illumination. DNA sequencing was performed by the MilleGen sequencing service (Labège, France).

1.4 Performance of bacterial cultures during fermentation in milk

1.4.1 Acidification kinetics of milk hydrolysates fermented by lactic acid bacteria at 37°C

Skimmed cow's milk (UHT) was inoculated with the selected lactic acid bacteria. Several aliquots were prepared to study the acidification kinetics of the milk by assessing the pH using a pH meter (Kika Laboretechnik pH meter) during 300 min of fermentation at 37°C.

1.5 Study of proteolytic activity

1.5.1 Test on UHT skimmed milk

The different isolates were reactivated twice by mixing 50μl of the preculture with 950 μl of milk and incubated for 48h at 37°C. The mixture was diluted 1:10 (V/V) in sample buffer containing 4% sodium dodecyl sulphate (SDS), 3% 2-mercaptoethanol, 20% glycerol, 50mM Tris-Hcl, pH 6.8 and heated at 100°C for 3 min, then analysed by electrophoresis (SDS-PAGE).

1.5.2 Proteolytic activity of lactic acid bacteria determined on sodium caseinates and denatured whey

Isolates that tested positive on UHT skimmed milk were cultured on Milk-Citrate-Agar (MCA) medium for 48 h (Fira et al., 2001) containing: 4.4% skimmed milk powder, 0.8% Na-citrate, 0.1 yeast extract, 0.5% glucose and 1.5% agar. Cells were collected and resuspended at $OD_{600=10}$ in 100mM Na phosphate buffer$^+$ at pH 6.8. Cells were added to a solution of sodium caseinates (12mg/mL) and denatured whey (85°C/20min) (5mg/mL) and incubated at 37°C for 3, 6, 9, 48h. At the end of the incubation period, centrifugation was carried out at 8,000 rpm for 10 min. The clear supernatant was collected and analysed by SDS-PAGE and HPLC.

1.5.3 Effect of pH and temperature on the proteolytic activity of lactic acid bacteria

The bacterial cells were recovered according to the protocol described above and placed in phosphate buffers at pH 5.4 and pH 6.6, then incubated at 37°C and 40°C for 48 hours.

1.5.4 Effect of inhibitors on bacterial proteolytic activity

To determine the protease type, the cells were cultured on MCA medium. After 48 h incubation, the cells were collected and re-suspended at $OD_{600=10}$ in phosphste buffer pH 7.2. We tested various proteolysis inhibitors (EDTA: Ethylene Diamine Tetra Acetic, a metallo-protease inhibitor, iodoacetic acid to inhibit cysteine proteases, and PMSF:

Phenylmethylsulphonyl fluoride, a serine protease inhibitor). These inhibitors were added to cell suspensions (10 OD600) at a final concentration of 10 mM, and incubated for 1 h at 37°C before adding the substrate. The samples were centrifuged for 10 min at 10,000 rpm and the cells were then mixed with a solution of sodium caseinates at a concentration of 12 mg/mL. Incubation was prolonged for up to 2 nights. The clear supernatant was recovered for analysis by SDS-PAGE to assess proteolytic activity.

1.5.5 Electrophoresis (SDS-PAGE)

Electrophoresis allows us to visualise bands corresponding to native proteins and peptides produced during fermentation. Electrophoresis was carried out using a Mini Protean II Gel Electrophoresis apparatus (Bio-Rad, Hercules, CA) according to the method of Schagger and Von Jagow (1988).

The composition of the gels is shown in Table 2. We used protein standards with molecular weights ranging from 14.4 to 97.4 kDa (Sigma). Fermented milk hydrolysate samples were diluted (v/v) in sample buffer (50mM Tris HCl pH 6.8, 4% SDS, 3% 2 β-mercaptoethanol, 10% glycerol and traces of bromophenol blue) and heated at 100°C for 3 minutes.

10µl of each sample was placed in each well of the separation gel. The separation was carried out using an electric field of 10mA in the concentration gel and 20mA in the separation gel.

The migration buffer used is : TRIS 50 mM, Glycine 0.384M, SDS 0.1%. After migration, the gels were stained with a mixture of ethanol 30%, acetic acid 5%, Coomassie blue R250 0.2% and H2O 65% for 1h and then decolourised (ethanol 30%, acetic acid 5%, H2O qsp 100%) for 1h.

1.5.6 High-performance reverse-phase liquid chromatography (HPLC)

This method enables highly resolutive separation of chemical or biological compounds thanks to a high pressure (several tens of bars) that eliminates the diffusion effects observed in atmospheric conditions.

To identify the chromatographic profile of peptide fractions from bacterial hydrolysis on sodium caseinates and denatured whey, reverse-phase liquid chromatography analysis was carried out using an HPLC (Waters 2695, Alliance) on a column (WATER Symmetry 300 C18, 2.1 x 150 mm).

The column was equilibrated with eluent A (0.055% TFA in H2O). Elution was carried out at room temperature with a linear gradient of 0-80% eluent B (80% CH3CN, 0.09% TFA in H2O). 10µl of sample was injected. Optical density was measured at 216nm.

Table 5: Composition of electrophoresis gels (12% SDS-PAGE)

Solutions	Concentration gel (Stacking gel)	Separation gel (Main freeze)
Acrylamide 40%	0.2 ml	1.8 ml
2M Tris pH 8.8		1.00 ml
0.5M Tris pH 6.8	0.3 ml	
Water	2 ml	3.2 ml
10% SDS	25 µl	60 µl
APS 10%	20 µl	60 µl
Temed	4 µl	12 µl

1.5.7 Determination of total protein in fermented milk

The concentration of residual proteins in the various hydrolysates of milk fermented for 48 hours was assessed using a kit, Bc Assay Protein Quantitation Kit, on a microplate at a wavelength of 540-590 nm using a spectrophotometer. The concentration was determined by reference to a standard range of bovine serum albumin.

1.6 Separation of peptide fractions by adsorption chromatography

In order to obtain hydrophobic peptide fractions of the hydrolysates after 48h of incubation at 37°C. The various hydrolysates were diluted in 1/1 (V/V) with 9M urea and centrifuged at 10,000 rpm for 10min at 4°C. The supernatant was then filtered through a column (sep-pack TC18 vac cartridge 3cc 37-55 µm Particle Size, 50/pK (WAT036815), made up of small silica particles and represents our solid phase. The column is placed on a vac elut Manifold connected to a pump with a maximum pressure of 15mmHg. After equilibrating the column, we first eliminated the hydrophilic fraction with Buffer A (0.03% Trifluoroacetic acid (TFA)). The hydrophobic fraction was eluted with Buffer B (80% acetonitrile, 1% iso-propanol, 10% Buffer A). The solvents in the recovered fractions were removed in a Speed Vac evaporator (Plus sc110A).

1.7 Determination of the antioxidant activity of the peptide fractions of different fermented milk hydrolysates

The antioxidant activity of the peptide fractions of different fermented milks was assessed by the Colorimetric technique using Trolox (6-hydroxy-2, 5, 7, 8-tetra- methylchroman-2-carboxylic acid, Sigma) (Re et *al*, 1999; Salami et *al*, 2011). This assay is based on the ability of an antioxidant to stabilise the cationic radical $ABTS^+$ (2,2'-azino-bis- (3-ethyl-benzthiazoline-6-sulfonic acid) supplied by Sigma-Aldrich. The initially blue-green coloured ABTS can be transformed into colourless $ABTS^+$ by the trapping of a proton by the antioxidant. A comparison was made with the ability of Trolox (a water-soluble structural analogue of vitamin E) to capture ABTS+. The decrease in absorbance caused by the antioxidant reflects the free radical capture capacity. The result is given in µM of Trolox equivalent per g of product. ABTS was used after oxidation according to the following protocol: 7mM of ABTS was added to 2.45 mM of Potassium Persulphate buffer, incubation lasted approximately 12 to 16 hours in the dark to induce the formation of the ABTS radical. The ABTS+ solution is diluted in 5mM Phosphate Buffer pH 7.4 to give an absorbance of 0.7 ±0.2 at a wavelength of 734nm. ABTS+ was added to the peptide fractions. The mixture obtained was incubated at 25°C for 6 min and the absorbance was measured at 734nm using a Shimadzu UV Visible Spectrophotometer 1800. Antioxidant activity is assessed with reference to a Trolox Standard curve (Trolox Equivalent Antioxidant Capacity) corresponding to the concentration of Trolox with the same activity as the test substance.

1.8 Inhibitory action of lactic acid bacteria against indicator and/or pathogenic bacteria

The various isolates were reactivated by depositing 50µl in 950 µl of SRM and incubated at 37°C overnight. The antimicrobial activity of the bacteria used was determined by the agar well diffusion method according to Schillinger and Lucke (1989). The test was carried out using bacterial cultures and their supernatant obtained after centrifugation at 10,000 rpm at 4°C for 15 minutes. The pH of the supernatant was adjusted to 6.5 with 1N NAOH, then filtered using a filter (Acrodisc filters, Pall, polyether sulphone, pore size 0.22 µm). As inhibitor strain we used *Escherichia coli* ATCC 23355 (American Type + culture collection), previously reactivated in nutrient medium overnight, then added to 20 ml nutrient agar (0.8%, w v). Wells were dug on the agar. We plated 100 µl of bacterial cultures and overnight supernatants into the wells. Incubation took place at 37°C for 18 hours. The presence of a clear zone of inhibition at least 2 mm in diameter was considered a positive result.

After obtaining a positive result and in order to determine the nature of the inhibiting agent, we treated the samples with catalase and proteinase K in phosphate buffer (0.1M, pH 7.0), under acidic conditions (pH 4.5), then under acidic conditions and neutralised at pH 6.5.

The enzymes were added to 200 µl of cell culture or supernatant at a final concentration of 0.1 mg/mL. After 2 h of incubation, the reaction was inactivated at 100°C for 3 min. Untreated cell cultures and MRS medium with catalase and proteinase K were used as controls.

1.9 Growth of lactic acid bacteria in MRS medium at different pH and bile concentrations.

1.9.1 Effect of pH on bacterial growth

The bacteria were cultured for 3 h in liquid SRM adjusted to pH 3, 4, 5, 6, 7, 9, 11 and 13 with 1N HCL or 1M NAOH, at a rate of 100 µl of SRM plus 5 µl of each bacterial culture. The cultures were then placed in MRS agar at pH 6.8 for 48 hours for bacterial enumeration on plates. The growth rate under both conditions was evaluated in colony-forming units per millilitre (CFU/ml).

1.9.2 Effect of bile on bacterial growth

Strains were cultured for 4h in liquid SRM containing 0.2, 0.3, 0.5, 0.6, 1.0, 2.0 and 3.0% bile (sigma) at a rate of 5 µl of each bacterial culture in 100µl of bile. Cultures in SRM without bile were used as controls. At the end of this operation, the bacteria were cultured in SRM agar pH 6.8 for 48 h and counted on plates. The growth rate in both conditions was evaluated in colony-forming units per millilitre (CFU/ml).

1.10 Assessment of antibiotic resistance

The antibiotic resistance of the strains studied, *E.faecium*, *E.faecalis* DAPTO 512, *L. paracasei*, and *L. plantarum* was evaluated by the disk diffusion method. The strains were regenerated in liquid SRM for 48 h to obtain a bacterial concentration of 10^9 cfu/ml. Each culture was swabbed onto an MRS agar plate. The antibiotics used were : Penicillin (P), Ampicillin (AM), Chloramphenicol (C), Ciprofloxacin (CIP), Gentamycin (GEN500), Vancomycin (VA), Kanamycin (K), Tetracycline(TE) (powder, Sigma). The antibiotic concentrations used ranged from 0.2 to 512 µgZml. The level of antibiotic sensitivity was reported as resistant (R), sensitive (SS) or intermediate (I) according to the sensitivity threshold recommended by the National Committee for Clinical Laboratory Standards (NCCLS).

2.   Evaluation of the therapeutic/preventive effect of cow's milk hydrolysates fermented by selected lactic acid bacteria

2.1 Assessment of the preventive effect

2.1.1 Measurement of the preventive effect of orally administered hydrolysates on the intestinal mucosa of Balb/c mice following intraperitoneal sensitisation to ß-Lg

2.1.1.1 Animals and rearing conditions

We used two hundred and ten female Balb/c mice (n=210) with an average weight of 18 to 22 g, aged 3 to 5 weeks, obtained from the Institut Pasteur in Algiers. These are female mice, bred and acclimatised before any manipulation in the animal house of the Nutrition Physiology and Food Safety Laboratory under housing conditions that comply with regulations. Experiments are carried out with respect for the animal's well-being, avoiding stress and agitation. The animals live in cages equipped with feeding bottles and a trough, are given tap water and fed *ad libitum* with rodent food obtained from SARL la Production Locale Bouzareah (Algiers).

The mice were divided into 07 groups of 10. The animals received oral hydrolysates of milk fermented by the different strains according to the following protocol:

Mice were given 0.2 ml of fermented milk hydrolysates orally for 18 days and then sensitised intraperitoneally to β-Lg. Sensitisation began on day 18^{éme} .

1. Negative control group: receives an oral solution of NaCl 9‰ at a rate of 0.2 ml for the duration of the experiment.

2. Positive control group: receives oral 9‰ NaCl solution at a rate of 0.2 ml for 15 days and then immunised with β -Lg intraperitoneally.

3. Group LF *E. faecium* receiving hydrolysates of milk fermented by *E. faecium* and then sensitised intraperitoneally to β -Lg

4. Group LF *E. faecalis* receiving hydrolysates of milk fermented with *E. faecalis* DAPTO 512 then sensitised intraperitoneally to β -Lg

5. LF *L. paracasei* group receiving hydrolysates of milk fermented by *L. paracasei* and then intraperitoneally sensitised to ß-Lg

6. Group LF (*Bifidobacterum longum- L. plantarum*) (Blg-Lp) receiving milk hydrolysates fermented with (*Blg-Lp*) then sensitized intraperitoneally to β-Lg

7. Group LF (*Streptococcus thermophillus- L. plantarum*) (*Strp-Lp)* receiving hydrolysates of milk fermented with (*Strp-Lp)* then sensitized intraperitoneally to β-Lg

2.1.1.2 Protocols for intraperitoneal immunisation with (ß-Lg)

The mice received 100 µl of a pH 7.4 PBS solution containing 10 µg of ß-Lg mixed with 2 mg of Al(OH) ₃. Intraperitoneal injections were given on D0, followed by repeat injections under the same conditions on days 14ème , 21ème and 28ème of the protocol.

2.1.2 Measurement of the preventive effect of orally administered hydrolysates on the intestinal mucosa of Balb/c mice following oral sensitisation to cow's milk

The mice were divided into 07 groups of 10. The animals received oral hydrolysates of milk fermented by the different strains according to the following protocol:

The mice were given 0.2 ml of fermented milk hydrolysates for 18 days and then orally sensitised to cow's milk to which 0.03 ml of Maalox had been added. Sensitisation began after day 18^{éme} of the experiment, every day during week 1ère , every other day during week 2ème and every third day during week 3ème .

1. Negative control group: receives an oral solution of NaCl 9‰ at a rate of 0.2 ml for the duration of the experiment.

2. Positive control group: receives an oral solution of NaCl 9‰ at a rate of 0.2 ml for 15 days and then sensitised orally to cow's milk.

3. Group LF *E. faecium* receiving hydrolysates of milk fermented by *E. faecium* and then orally sensitised to cow's milk.

4. Group LF *E. faecalis* receiving hydrolysates of milk fermented by *E. faecalis* DAPTO 512 and then orally sensitised to cow's milk.

5. Group LF *L. paracasei* receiving hydrolysates of milk fermented by *L. paracasei* and then orally sensitised to cow's milk.

6. Group LF (*Blg-Lp)* receiving hydrolysates of milk fermented by (*Blg-Lp)* and then orally sensitised to cow's milk.

7. Group LF (*Strp-Lp)* receiving hydrolysates of milk fermented by (*Strp-Lp)* and then orally sensitised to cow's milk.

2.2 Evaluation of the therapeutic effect

The mice were divided into 07 groups of 10 and were intraperitoneally sensitised to ß-Lg and then given fermented milk hydrolysates.

1. Negative control group: receives an oral solution of NaCl 9‰ at a rate of 0.2 ml for the duration of the experiment.

2. Positive control group: immunised with ß-Lg intraperitoneally and given 0.2 ml NaCl 9‰ solution orally for 15 days.

3. Group LF *E. faecium* sensitised intraperitoneally to β-Lg and then given *E. faecium* fermented milk hydrolysates

4. Group LF *E. faecalis* sensitised intraperitoneally to β-Lg and then given hydrolysates of milk fermented with *E. faecalis* DAPTO 512

5. Group LF *L. paracasei* sensitised intraperitoneally to ß-Lg and then given *L. paracasei* fermented milk hydrolysates

6. Group LF (Blg-Lp) then sensitised intraperitoneally to β -Lg then given hydrolysates of milk fermented by *(Blg-Lp)*

7. Group LF *(Strp-Lp)* sensitised intraperitoneally to β-Lg and then given hydrolysates of milk fermented by *(Strp-Lp)*.

2.3 Blood sampling

Blood samples were taken at the end of each experiment, before the animals were killed.

Blood was collected by retro-orbital puncture using a Pasteur pipette and centrifuged at 3,500 rpm for 15 min at 4°C. Sera were aliquoted and stored at -20°C for immunoglobulin assay.

2.4 Assessing the degree of animal awareness

2.4.1 Determination of serum anti-B-Lg IgG and IgE

In order to assess the degree of sensitisation of the animals in all groups of mice, serum anti-B-Lg IgG and IgE were titrated by an enzyme-linked immunosorbent assay (ELISA) using a non-competitive procedure inspired by the technique of Engvall & Perlmann (1971).

2.4.1.1 Principle of indirect colorimetric ELISA

The principle of this technique is based on a process in which the antibodies to be measured first react with the immobilised antigen by adsorption onto a solid phase. In a second step, the quantity of antibodies bound by the antigen is measured using a second antibody (anti-immunoglobulin).

In our work, we use polystyrene microtiter plates (NUNC MaxiSorp, flat-bottomed wells), which allow most antigens to be adsorbed. Once the immunoserum containing the specific antibodies has been deposited, the solid phase is washed and the presence of these antibodies is revealed by the addition of a conjugate corresponding to anti-antibodies that may or may not be coupled to an enzyme (peroxidase). The final step is the assay of the marker enzyme. This phase is essential, because the sensitivity threshold depends on the smallest number of enzyme molecules that can be detected.

The peroxidase substrate we use is hydrogen peroxide (H_2O_2).

During the enzymatic reaction, an O radical is formed. To detect this, a chromogen, orthophenylene diamine (OPD), is added to the solution.

How it works

The ELISA assay for specific anti-BLG IgG and IgE is carried out by the indirect ELISA technique using 96-well flat-bottom plates (NUNC MaxiSorp), according to the following steps:

> All microplate wells receive 100 µl of antigen at the concentration of 10 µg/ml of ß-Lg, diluted in PBS pH 7.4. The plates are then incubated for at least overnight at +4°C.

> Excess unfixed antigen is removed by washing the plate 3 times with 0.05% PBS-Tween 20 using an automatic washer (Elx50).

> The non-specific sites were saturated by depositing 200 µl of 3% SAB in PBS pH 7.4 in all the wells.

> The plate was then incubated for 1 hour at 37°C and rinsed 3 consecutive times with shaking using 0.05% PBS/Tween 20 wash buffer.

> The next step is to dilute the serum samples to be tested to 1:10ème in dilution buffer (PBS 0.01M/ SAB 1% Tween 20 0.1% pH 7). To do this, a series of dilutions ranging from 10^{-1} to 10^{-7} is then performed. A volume of 100 µl was then added to the appropriate wells.

> The plate was then incubated at 37°C for 2 hours and washed 3 consecutive times with shaking, using 0.05% PBS/Tween 20 wash buffer.

> Each well of the plate then receives 100 µl of mouse antibody diluted to 1/20000ème in dilution buffer, depending on the antibodies being tested for. The antibody deposited is either anti-IgG or anti-IgE (Sigma).

> The plate was then incubated for 1 hr 30 min at 37°C, followed by 3 consecutive washes with PBS/Tween 20 0.05% wash buffer.

> Next, 100 µl of extravidin peroxidase (Sigma) diluted 1:5000ème in pH 7 dilution buffer was added to each well. The plate was then incubated for 30 min at 37°C.

> After washing with 0.05% PBS/Tween 20, 200 µl of a solution containing a chromogen (orthophenylene diamine (OPD): 6 mg diluted in 20 ml of 0.05M sodium citrate buffer at pH 5.1 and 30 µl of H$_{2}$0$_{2}$) was added to each well of the plate. A colour reaction developed in 30 minutes at room temperature and protected from light. The addition of 2N H2SO4 stopped the reaction.

> The intensity of the colorimetric reaction is measured at 492 nm using a reader (ELx800).

> Positive and negative controls were included in each plate to check the specificity and sensitivity of each measurement.

Table 6: Composition of saline phosphate buffer (1 M), pH=7.4

Solution	Quantity
Na2HPO4, 2H2O	29 g
KH2PO4	2 g
NaCl	80 g
KCl	2 g
Thymerosal	1 g
Ultra pure water	1000 ml

Table 7: Composition of ELISA buffer solutions.

Tampons	Products
Capture pad	NaHCO3 (0.1 M) pH 9.6
Washing pad	PBS (0.01 M) pH = 7.4 Tween 20 0.05
Saturation buffer	PBS (0.01 M) pH = 7.4 BSA 3%
Dilution buffer	PBS (0.01 M) pH = 7.4 BSA 3% Twen 20 0.1

Table 8: Composition of citrate buffer pH=4

Solution	Preparation method
Citrate buffer pH=4	294 mg citrate + 472 µl acetic acid + H2O until pH=4 is reached

Table 9: Composition of the Orthophenylene diamine

(OPD) developer solution

Solution	Composition
Developer solution (OPD)	**60mg OPD + 20 ml citrate buffer pH=5.1 + 30 µl H2O2.**

2.4.2 *Ex vivo* gut challenge test in Ussing chamber

This test was applied in groups of mice sensitized intraperitoneally to β- Lg

2.4.2.1 Principle of the Ussing chamber

The Ussing chamber (Schwan, 1968; Li, 2004) can also be used to measure epithelial electrical parameters such as short circuit current (*Isc*), resting transepithelial potential difference (*DDP*) and membrane conductance (*G*). The Ussing chamber takes its name from its creator, H.H. Ussing, who in 1951 published a method for studying the flow of $Na+$ ions through a frog's skin and was thus able to determine the role played by this ion in membrane potential. His method of investigation consisted in cancelling the potential difference on either side of the membrane, so as to eliminate the passive transport of ions and observe only the net active flow, an image of the current resulting under short-circuit conditions. This technique was adapted to the intestine by Stanley and Zalusky (1964). Since then, this method has been applied to numerous laboratory animal models and is used for fragments of human intestine removed during surgery or intestinal biopsies (Saidi *et al.*, 1995). The fragment of intestinal mucosa can be mounted flat between two lucite half-chambers whose opening, determining the exposed surface, is adapted to the size of the fragment to be studied (0.1 to 0.2 cm^2).

The Ussing chamber is basically made up of two compartments separated by the tissue to be studied. The two compartments containing a physiological solution (Ringer) are used to isolate the luminal and serous sides of the intestinal tissue in order to reproduce the principle of the selective barrier as closely as possible. The compartments are usually made of Plexiglas (polymethyl methacrylate), which is appreciated for its transparency and behaviour with respect to biological tissue. The physiological solution is maintained at a constant temperature of 37°C and is continuously oxygenated by bubbling carbogen (95% O_2 and 5% CO_2). The other effect of bubbling is to continuously stir the survival fluid. Under these environmental conditions, most tissues retain a functional viability of more than 2 hours, making it easy to study active and passive transport through the tissue.

The very principle of the Ussing chamber is that there should be no electrochemical gradient allowing a given solution to pass through the tissue. This condition is met when both sides of the tissue are bathed in solutions of the same chemical composition, temperature, pH and osmolarity.

In addition to conventional permeability measurements, the Ussing chamber coupled to clamps can be used to check the state of the membrane by measuring the electrical parameters *Isc*, *Vm* and *G* (conductance). For these measurements, the Ussing chamber behaves like an electrochemical cell with its stimulation and measurement electrodes. The stimulation electrodes are used to impose the current in the chamber, which cancels out *Vm*.

2.4.2.2 Mounting of mouse jejunum fragments in the Ussing chamber

The mice were kept fasting overnight. Once anaesthetised, the abdomen was opened and the entire jejunal segment was gently removed from the abdominal cavity, its contents removed by rinsing several times with cold Ringer. The jejunal segment is then incised along the mesenteric border. It is then cut into fragments, which are kept in cold Ringer's and oxygenated with a stream of carbogen (CO_2: 5%, O_2: 95%).

Two millilitres of Ringer's are deposited in the two compartments of the chamber maintained at 37°C and oxygenated by a continuous stream of carbogen and after 30 minutes of stabilisation of the electrophysiological parameters. 20 µg/ml of ß-Lg is deposited in the serous compartment. The various electrophysiological parameters were then measured initially every minute for the first five minutes, then once every 5 min, for 15 min of the experiment. The mucosal and serosal compartments were connected to a pair of calomel electrodes via agar bridges (4 g/100 ml 3M KCl) on both sides of the tissue, enabling the spontaneous potential difference (PD) of the tissue (serosal positive) to be measured. It is possible to suppress this PD and bring it to 0 mV, using a system of Ag/AgCl electrodes connected to a current source that allows a low-intensity current (10 µA) to flow through the tissue. This current is called the short-circuit current (Isc): it represents the sum of the net ionic fluxes, mainly Na^+ and Cl^- and a residual flux of HCO ions^{-3} , **Isc = J^{Na+} net + J^{Cl-} net + Jr.**

Ohm's law (U=RI) can be applied to determine the resistance of the fluid (Rf) in the absence of the fabric and the resistance of the fabric (Rt) once it is fitted in the chamber, or its inverse: the conductance G (G=1/R=1/U).

Table 10: Composition of Ringer's solution

Na^+	140mM
K^+	5.2mM
Ca^{++}	1.3mM
Mg^{++}	1.2 mM
CL-	120mM
$HCO3^-$	25mM
HPO4	2.4mM
H2PO4	0.4 mM

2.5 Histological study

The aim of this study was to check whether there were any changes in the histological structure of the intestinal epithelium, particularly in the jejunal villous architecture of mice in the different experimental groups compared with the negative control group.

The samples used undergo a number of stages beforehand:

2.5.1 Mounting

Tissues are fixed in 10% buffered formalin. Formaldehyde solutions are the most widely used fixatives.

2.5.2 Dehydration

After fixation, the tissues were dehydrated in 3 successive acetone baths, each bath lasting 45 minutes.

2.5.3 Clarification

After dehydration, the parts are placed in 2 xylene baths. Each bath lasts 45 minutes.

2.5.4 Inclusion

Inclusion is carried out using paraffin, which is a mixture of solid hydrocarbons with a high molecular weight and low affinity. These substances are characterised by their indifference to chemical agents. The samples are placed in two successive paraffin baths for one hour each at a temperature of 56°C and then cast in metal moulds. Plastic cassettes are then attached to the moulds, and the volume is topped up with paraffin before being placed in the freezer for 15 minutes to ensure solidification.

After paraffin embedding, the blocks containing the fragment are cut using a microtome to a thickness of 7 μm.

Once the sections have been made, they are placed on a glass slide covered with glue (2 g of albumin + 50 ml of glycerine in 1000 ml of distilled water) and then placed on a hot plate set at a suitable temperature, below the melting point of the paraffin. Using forceps, the folds of paraffin are pulled slightly to either side, then the cutter-slide assembly is removed from the plate, drained and wrung out with Joseph paper. Before staining, the slides are dewaxed and rehydrated.

2.5.5 Dewaxing

To dewax the slides, simply place them in two successive toluene baths. Each bath lasts 10 minutes.

2.5.6 Rehydration

It is carried out in 3 successive baths of ethyl alcohol of decreasing degrees (100°, 95°, 90°, 70°). Each bath lasts 2 minutes; the last is followed by a rinse under running water.

2.5.7 Colouring

Our slides were stained with haemalun-eosin, the simplest of the combined stains. A "basic" nuclear stain, haematein, and an "acidic" cytoplasmic stain, eosin, were used in succession. The nucleus stains blue-black and the cytoplasm pink to red. Preparation of the Harris haematoxylin dye (haematoxylin 5g, ethanol 50ml, potassium alum 100g, distilled water 1000ml, mercuric oxide 2.5g) (Hould, 1984).

Staining protocol

- Place the slides in Harris haematoxylin for 2 to 3 minutes.
- Wash the slides in normal water for 5 minutes.
- In the event of excessive staining, the slides are soaked in hydrochloric alcohol for a few seconds (100 ml of 95° alcohol + 5 drops of 1% Hcl).
- rinsing in a saturated aqueous solution of lithium carbonate.
- Wash the slides in ordinary water.
- Place the slides in an ethyl alcohol bath for 1 to 2 minutes.
- Stain slides with alcoholic eosin (2g eosin in 100ml ethyl alcohol) for 5 minutes.
- The slides were rinsed in two successive baths of 70° and 95° ethyl alcohol.
- Place the slides in toluene for 1 minute.
- Put a drop of Canada balsam or eukitt between the slides and the coverslip and leave to dry, then observe under the light microscope.

2.6 Statistical study

Results are presented as mean ± standard error. Means are compared using Student's t-test. The significance level used is that usually considered, i.e. 5%.

CHAPTER 6

Results and discussion

1. Properties of cow's milk hydrolysates after fermentation by lactic acid bacteria

1.1 Isolation of lactic acid bacteria

Six strains of lactic acid bacteria were selected for this study.

After examination under the light microscope, the bacteria were classified as 02 bacilli, 03 coccus and 01 bifidobacterium (table 11). They were gram positive and catalase negative and the *Bifidobacterium* strain was catalase positive. These observations enabled us to classify the isolates according to Gram, cell morphology and mode of association (Joffin et Leyral, 1996). Our results show a predominance of shells compared to rods. These results are in line with those obtained by El Soda et *al*, (2003) and El-Baradei et *al*,(2008) and El- Ghaish et *al*,(2011b) in different dairy product samples with a predominance of shells.

1.2 Biochemical identification of lactic acid bacteria

After fermentation of the carbohydrates observed on the API 50CH gallery and API strept gallery, we were able to identify the following species: *Lactobacillus paracasei (L. paracasei), Lactobacillus plantarum (Lp), Enterococcus faecium (E. faecium), Enterococcus faecalis (E. faecalis), Streptococcus thermophillus (Strp).*

These lactic acid bacteria were stored at -20°C in 15% SRM and 30% glycerol (Merck) for subsequent experiments. The micro-organisms were activated in 05 ml sterile aliquots of broth (SRM) (Merck) at 37°C for 24 h.

These lactic bacteria have been used to prepare fermented milk hydrolysates.

Table 11: Morphological characteristics of lactic acid bacteria isolated from cow's milk

Features	Lactic acid bacteria				
	1	2	3	4	5
Gram	G+	G+	G+	G+	G+
Cellular morphology	hull	hull	bacillus	bacillus	hull
Catalase	-	-	-	-	-

1.3 Identification by 16S DNA *PCR*

Strains were purified on MRS medium. DNA extraction was performed using the Kit (Dneasy blood & tissue).

One strain was amplified by 16sDNA. A product of 15000 (bp) was observed on the *agarose* electrophoresis gel (fig.10). After sequencing (MilleGen, Labège, France), one strain was identified as *Enterococcus faecalis* DAPTO 512. The remaining four strains are currently being identified.

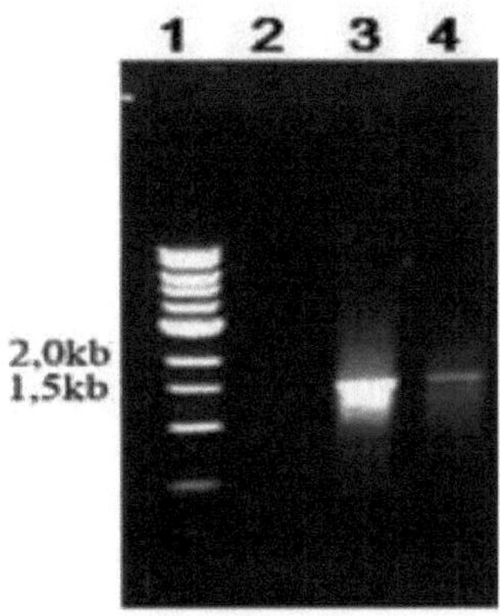

Fig. 10 PCR products obtained by amplification of the 16S rDNA gene.

Well 1. Molecular weight marker (GeneRuler™ 1kb DNA Ladder);
Well 2: negative control (contains no DNA);
Well 3. positive control ;
Well 4: amplified DNA fragment.

1.4 Performance of bacterial cultures during milk fermentation 1.4.1 Acidification kinetics of milk hydrolysates fermented by lactic acid bacteria at 37°C

After reactivation of the selected bacteria in 05 ml sterile aliquots of MRS broth at 37°C for 24 h, they were inoculated into UHT skimmed milk and incubated at 37°C for 300 min. They were inoculated into UHT skimmed milk and incubated at 37°C for 300 min.

The results (fig.11) show a difference between the pH of the different hydrolysates of fermented milk compared with control milk after 18h of fermentation. The pH of the control milk was stable throughout the experiment (6.59±0.83). However, the pH values of the different hydrolysates decreased significantly during fermentation.

Acidification kinetics are a good index of fermentation, and our results show that all of our strains have an interesting fermentation profile, since we observed the start of acidification as early as the first hour. This result is in line with that of Vrancken et *al* (2008). In his study, he investigated the growth and consumption of sugar, the production of acetic and lactic acid and the production of mannitol by *L. fermentum* IMDO 130101, and showed that maximum bacterial density was obtained between pH4 and pH7. This fermentation profile was also observed by Amiot et *al* (2002), which shows an acidification of the medium accompanied by physical, chemical and organoleptic changes.

Our results also indicate, as reported by El-Ghaish et *al.* (2010), that milk is a favourable medium for the growth of lactic acid bacteria.

After 300 min of fermentation we observe that the lowest pH is obtained in LF *L. paracasei* with a pH of 5.07±0.44, followed by LF *E. faecalis, LF L. plantarum-B. longum (Lp-Blg), LF Streptococcus thermophillus-Bifidobacterium longum (Strp-Blg) and LF*
E. faecium with pH values (5.3±0.40, 5.52±0.43, 5.38±0.44, 5.56±0.31) respectively. As indicated by Freitas et *al* (1999), our results show that the level of acidification varies according to the species studied. Indeed, he showed that *E. faecium* and *E. faecalis* strains

degrade ovine and caprine lactose more slowly than *L. paracasei*. In the same context, a study carried out on several strains of *E. faecium* and *E. faecalis* on growth and acidifying power showed that the highest acidifying activity was observed in *E. faecalis* (Villani and Coppola, 1994; Suzzi et *al.*, 2000).

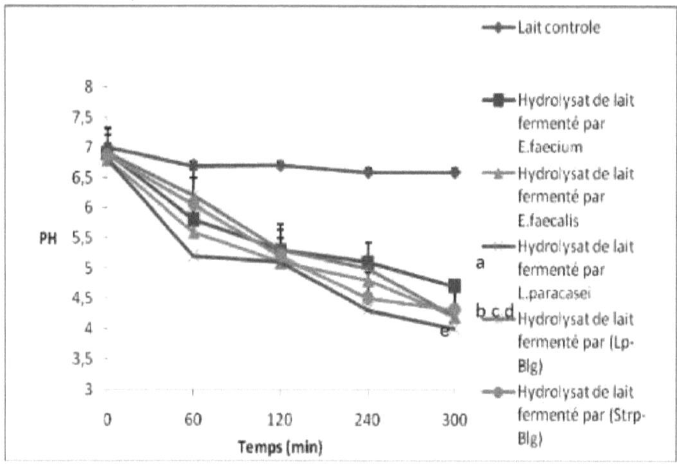

Fig. 11 Acidification kinetics of milk hydrolysates fermented by lactic acid bacteria at 37°C.

Values are presented as X±SE (n=3);

The (*p*) values represent comparisons of the average pH of the different hydrolysates with the control milk.

[a]*p* hydrolysate of milk fermented by (E. *faecium*) *p<0.01;*[b] p hydrolysate of milk fermented by *Streptococcus thermophillus-B. longum (Strp-Blg) p<0.01;*[c] p hydrolysate of milk fermented by L. plantarum-Bifidobacterium *longum* (Lp-Blg) *p<0.01;*[d] p hydrolysate of milk fermented with E. faecalis *p<0.01;*[e] p hydrolysate of milk fermented with *L. paracasei p<0.001*

1.5 Study of the proteolytic activity of lactic acid bacteria
1.5.1 Test on UHT skimmed milk

The lactic acid bacteria were tested to determine their proteolytic activity on skimmed UHT milk after 48h at 37°C . The various isolates were reactivated twice by mixing 50µl of the pre-culture with 950 µl of UHT milk, followed by incubation for 48h at 37°C, then analysed by electrophoresis on polyacrylamide gel (SDS-PAGE).

Examination of the electrophoretic profiles (fig.12) shows that the 3 strains have different degrees of proteolytic power after 48h of incubation. The highest hydrolysis power was observed in the *E. faecium* strain, followed by *L. paracasei* and *E. faecalis* for milk caseins.

In the case of whey proteins, the most marked hydrolysis of ß-Lg was observed in the *E. faecalis* strain, with the production of peptides with MW <14KDA, followed to a lesser extent by *E. faecium*, which has a low hydrolysis capacity (fig.12).

These results are in agreement with those of (Franz & Hozalpfel, 2003) and (El-Ghaish et *al.* 2010) who show that enterococci are the most proteolytic cocci and this property confers sensory qualities on dairy products. This explains why the majority of the enterococcus genus is present in the Egyptian agri-food industry (El-Ghaish et *al.* 2010). El Soda et *al* (2003) confirmed this idea by stating that *E. faecium* and *E. faecalis* strains are the most selected bacteria with a percentage of 32% and 3% respectively.

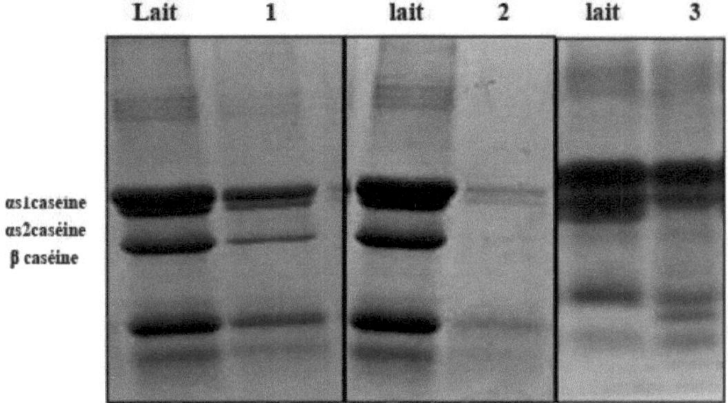

Fig. 12 SDS-PAGE electrophoretic profile of milk hydrolysates fermented by lactic acid bacteria after 48h of fermentation at 37°C.

Well1. Fermented milk hydrolysate (LF *L. paracasei*);

Well 2: Fermented milk hydrolysate (LF *E. faecium*) ;

Well 3: fermented milk hydrolysate (LF *E. faecalis DAPTO 512*).

1.5.2 Proteolytic activity determined on sodium caseinates and denatured whey

1.5.2.1 Electrophoresis analysis of milk hydrolysates fermented by two cocultures *Streptococcus thermophillus- L. plantarum* (*Strp-Lp*) and *Bifidobacerium longum-L. plantarum* (*Blg-Lp*).

The proteolytic activity (fig.13) of the two milk hydrolysates fermented by the two mixed cultures shows that (*Strp-Lp*) hydrolysed sodium caseinates, the αs2 and β casein fraction was hydrolysed with release of peptides with a molecular weight < 35KDA.

In the mixed culture (*BLg-Lp*), hydrolysis appears to affect only β-casein, with the release of peptides with a molecular weight < 35KDA. Our results show that the two associations we tested do not have the same proteolytic profile.

Shihata and shah, (2006), reported that good growth of bifidobacteria was observed when associated with a lactobacillus. Altieri et *al*, (2008) added that active synergies develop between lactic acid bacteria in mixed cultures. It has also been reported by (Fernandez-Espla et *al.*, 2000; Hols et *al.*, 2005; Liu et *al.*, 2010, Hafeez et *al.*, 2013) that the proteolytic system of *S.thermophillus* consists of extracellular proteases called PRT, an amino acid and peptide transport system and a set of intracellular peptidases.

1.5.2.2 Electrophoretic profile and HPLC chromatographic analysis of Caseinates of Sodium hydrolysed by *L. paracasei*

The proteolytic activity of the *L. paracasei* strain on sodium caseinates was observed from 3h of incubation; the αs1-αs2 and β-casein fraction were hydrolysed, and low molecular weight peptides <35 kDA were released. Proteolysis increased with incubation time to reach a maximum after 48h. Total hydrolysis of β-casein was observed with the disappearance of bands corresponding to the peptides resulting from this hydrolysis (fig. 14A).

The HPLC chromatographic profile (fig.14B) of the same sample showed that hydrolysis of sodium caseinates after 48h of incubation generated peptides, with retention times between 19 and 31 min of the elution time. The peptides were eluted at a 40% linear gradient corresponding to the hydrophobic zone of the mobile phase.

This hydrolysis capacity has also been observed by Kunji et *al* (1996) and Sarantinopoulos et *al* (2001). A study carried out on the strain *Lactobacillus fermentum*.

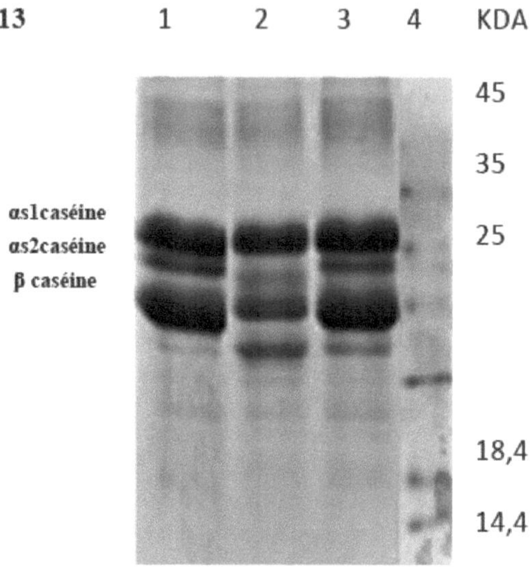

Fig.13 SDS-PAGE electrophoretic profile (12%) of sodium caseinates hydrolysed by the two co-cultures *Streptococcus thermophillus-L. plantarum* (Strp-Lp) *and Bifidobacterium longum- L. plantarum* (Blg-Lp) after 48h incubation at 37°C.

Well 1. Sodium caseinates without bacterial cells;

Well 2. Hydrolysates of sodium caseinates in the presence of (Strp-Lp) at different times at 37°C.

Well 3. Hydrolysates of sodium caseinates in the presence of (Blg-Lp) at different times at 37°C.

Well 4. peptide kit.

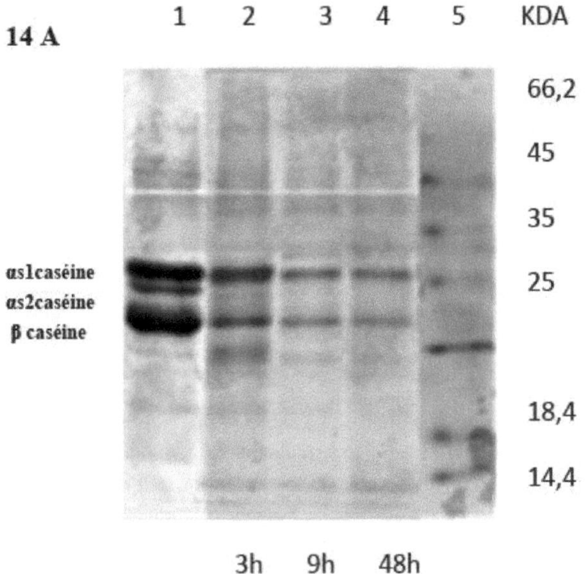

Fig.14 A SDS-PAGE analysis (12%) of sodium caseinate hydrolysates obtained after fermentation by _L. paracasei_ at 37°C after 3h, 9h and 48h.
Well 1. Sodium caseinate without bacterial cell ;
Wells 2, 3, 4. Hydrolysates of sodium caseinates ;
Well 5. peptide kit.

14 B

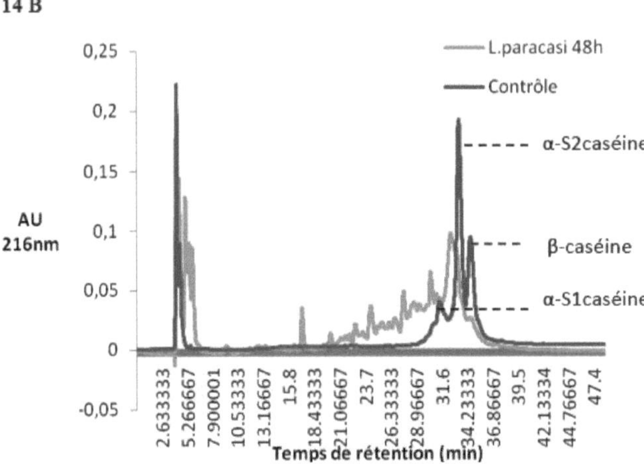

Fig.14 B HPLC chromatographic profile of peptide fractions from hydrolysis of sodium caseinates by _L. paracasei_ after 48h.

The retention time for sodium caseinate degradation products was between 19 and 31 min. The elution gradient was between 40 and 60%, corresponding to the hydrophobic zone of the mobile phase.

El-Ghaish et *al,* (2010) showed total hydrolysis of β and K caseins with peptide release. A similar study by Tzvetkova et *al,* (2007) also showed that lactobacilli isolated from three types of homemade yoghurt from the Balakan region hydrolysed between 80-90% of β-casein, with a lower degree for both α-S1 and a-S2 fractions. These results suggest that proteases from the *L. paracasei* strain can be classified as type PIII, as these proteases have the ability to hydrolyse both αs1 and β casein fractions (Kunji et *al.*, 1996; El-Ghaish et *al.* ,2010).

1.5.2.3 Electrophoretic profile (SDS-PAGE) and HPLC chromatographic analysis of Caseinates of Sodium and denatured whey hydrolysed by *E. faecium*

- *Sodium caseinates*

After 3 h of incubation, we observed significant hydrolysis of the a- S2 and β-casein fractions by the *E. faecium* strain, with the appearance of peptides with a molecular weight <35 k DA. This hydrolysis was virtually complete after 48 h (fig. 15C). According to (Kunji et *al.*, 1996) *E. faecium* proteases can be classified as PI types. A study by Arizcum et *al* (1997) showed that enterococci are the most predominant proteolytic bacteria.

The HPLC profile of sodium caseinates (fig. 15D) showed that hydrolysis generated peptides which eluted at a linear gradient of between 40% and 60% and appeared in the hydrophobic zone of the mobile phase, with retention times of between 1932 min.

- *Whey*

We observe a slight hydrolysis of β-Lg, but a-La does not seem to be affected by *E. faecium* (fig. 16 E).

The HPLC profile (fig. 16 F) showed that peptides resulting from the degradation of β-Lg appeared in the hydrophobic zone of the mobile phase. Their retention time is between 16-22 min corresponding to the 40% - 50 % linear gradient. These results are in agreement with those of El-Ghaish et *al,* (2010) who show a low proteolytic activity of enterococcus. In contrast, lactic acid bacteria such as *Lactobacillus acidophillus* CRL.636, *Streptococcus thermophillus* CRL.804 and *L.delbruekii spp bulgaricus* CRL 454 hydrolyse the majority of whey fractions (21% and 26% of β-Lg and a-La, respectively) (Pescuma et *al.*, 2008).

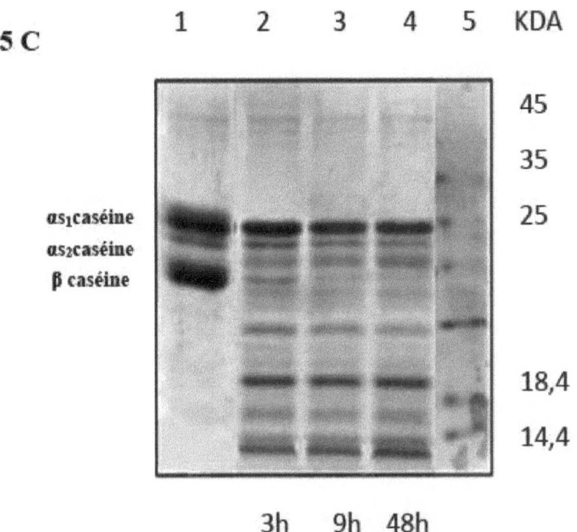

Fig.15 C SDS-PAGE analysis (12%) of sodium caseinates hydrolysed by *E. faecium* at different times at 37°C.

Well 1. Sodium caseinates without bacterial cells;

Wells 2, 3, 4. Hydrolysates of sodium caseinates in the presence of *E. faecium* at different times at 37°C;

Well 5. peptide kit.

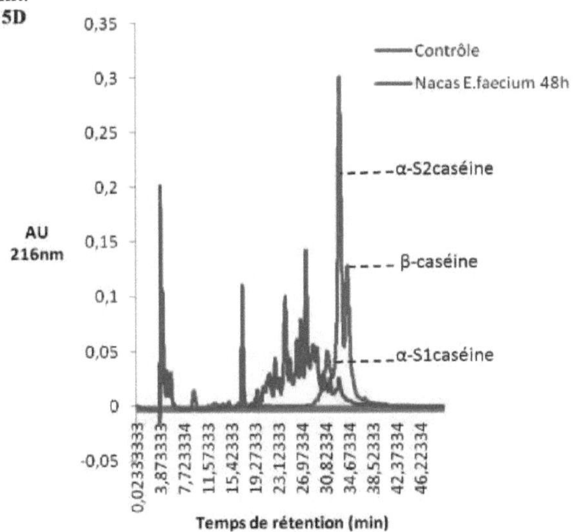

Fig.15 D HPLC chromatographic profile of peptide fractions from hydrolysis of sodium caseinates by *E. faecium* after 48h.

The retention time for sodium caseinate degradation products was between 19 and 32 min.

The elution gradient between 40% and 70% corresponds to the hydrophobic zone of the mobile phase.

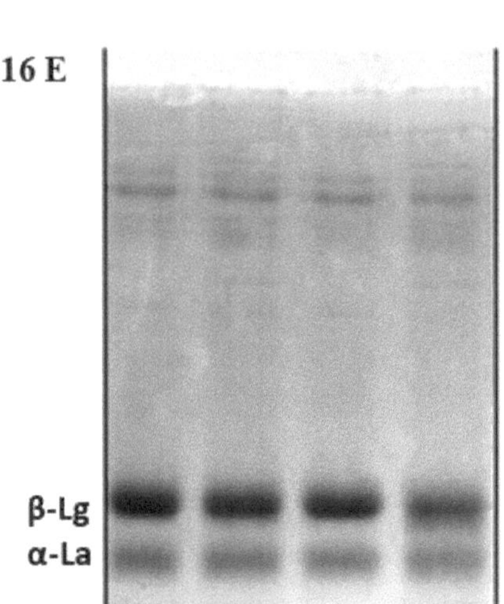

Fig. 16 E SDS-PAGE profile (12%) of whey denatured by *E. faecium* after 48h at 37°C.
Well 1. Whey without bacterial cells;
Wells 2, 3, 4. Hydrolysates of denatured whey in the presence of E. *faecium* after 48h at 37°C.

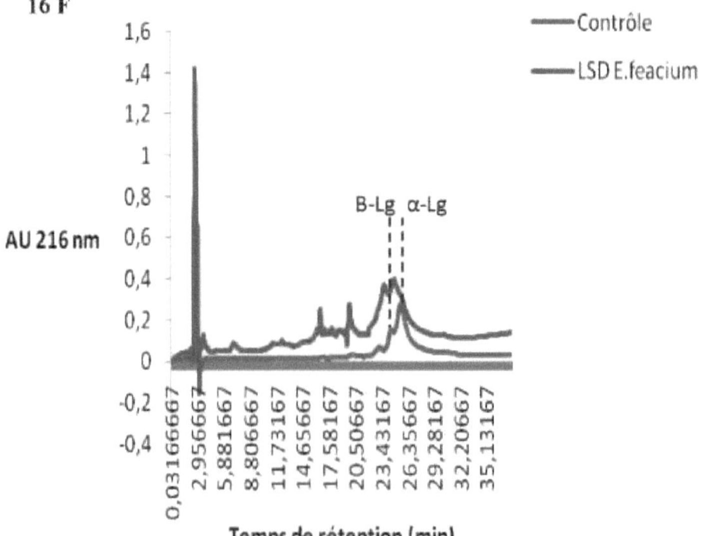

Fig.16 F HPLC chromatographic profile of peptide fractions from hydrolysis of whey

42

denatured by *E. faecium* after 48h.

The retention time for the degradation products of denatured whey was between 16 and 22 min. The elution gradient was between 30% and 50%, corresponding to the hydrophobic zone of the mobile phase.

1.5.2.4 Analysis by SDS-PAGE electrophoresis (12%) and HPLC chromatographic profile of Caseinates of Sodium and denatured whey hydrolysed by *E. Faecalis DAPTO 512*

-Sodium caseinates

The electrophoretic profile (fig. 17G) shows that *E. faecalis* strain DAPTO 512 is less proteolytic than the other strains. Hydrolysis is lower for sodium caseinates. After 9 h of incubation only the β-casein fraction was hydrolysed, with the appearance of peptides with a molecular weight <30 kDA and after 48 h of incubation hydrolysis was more advanced. According to the classification of Kunji et *al*, (1996), the *E. faecalis DAPTO 512* protease can be classified in the PI type.

The HPLC profile (fig. 17H) of sodium caseinates showed that hydrolysis of sodium caseinates after 48 h of incubation generated hydrophobic peptides eluted at a linear gradient of between 30% and 60% corresponding to the hydrophobic zone of the mobile phase. The retention time of these peptides was between 24-30 min.

- Whey

Our results show that *E. faecalis* strain DAPTO 512 degrades ß-Lg and not a-La after 48 h incubation. This proteolytic activity is evidenced by the appearance of peptides with a molecular weight <18kDA (fig. 18 I).

The HPLC profile (fig. 18 J) of the whey showed that hydrolysis of whey proteins by *E. faecalis* DAPTO 512 generated peptides. The retention time of these peptides is between 20-30 min. They eluted in the hydrophobic zone of the mobile phase at between 30% and 50%. These results are in agreement with those of El-Ghaish et *al* (2010). He showed that an *E. faecalis* strain hydrolysed only the ß-Lg of whey.

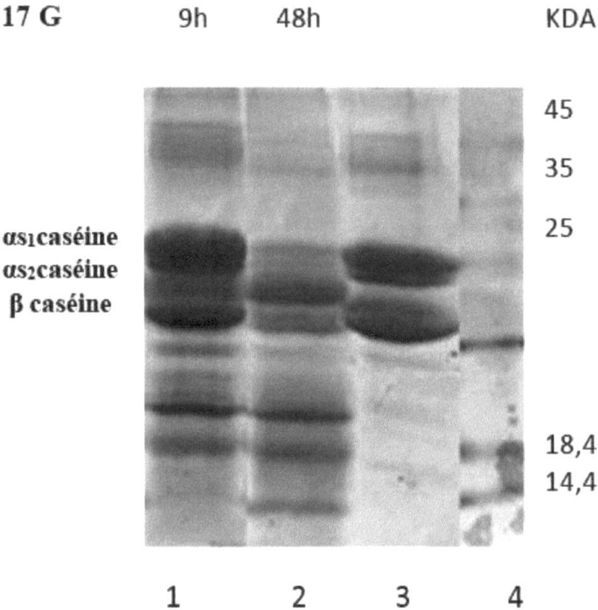

Fig. 17 G SDS-PAGE analysis (12%) of sodium caseinates hydrolysed by *E. Faecalis DAPTO 512* after 9 and 48 h at 37°C.

Well 1. Hydrolysates of sodium caseinates in the presence of E. *faecalis* after 9h;

Well 2: Hydrolysates of sodium caseinates in the presence of E. *faecalis* after 48h;

Well 3: Sodium caseinates without bacterial cells ;

Well 4. peptide kit.

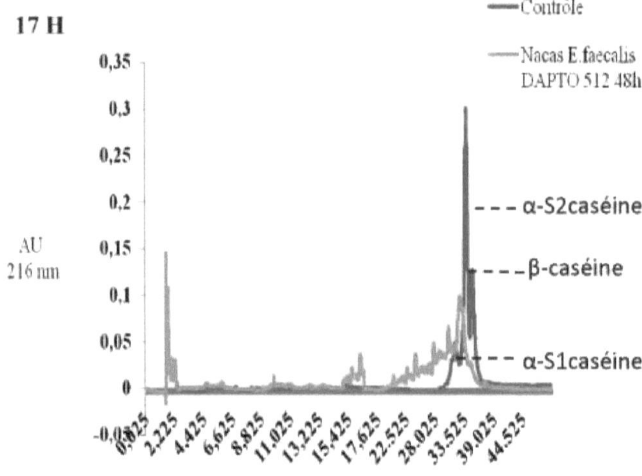

Fig. 17 H HPLC chromatographic analysis of peptide fractions from hydrolysis of

sodium caseinates by *E. faecalis DAPTO 512* after 48h.

Hydrophobic peptides are eluted at a linear gradient of between 30% and 60% of the mobile
phase, and the retention time for these peptides is between 24-30 min.

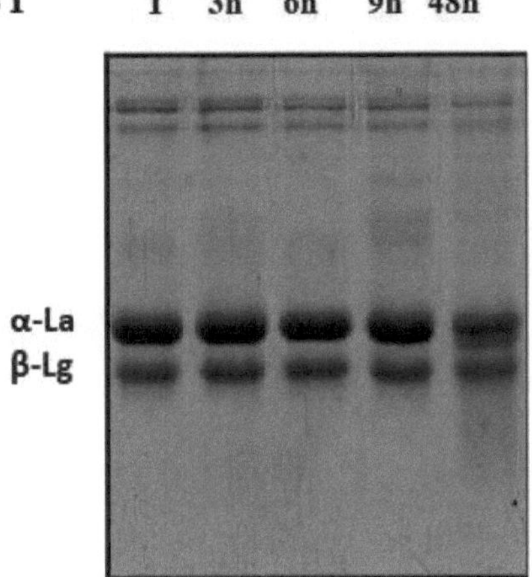

Fig.18 I SDS-PAGE analysis (15%) of denatured whey hydrolysed by *E. Faecalis*
***DAPTO 512* at different times at 37°C**
Well 1. Denatured whey without bacterial cells;
Wells 2, 3, 4, 5. Hydrolysis of denatured whey by *E. faecalis DAPTO 512* at different times at
37°C.

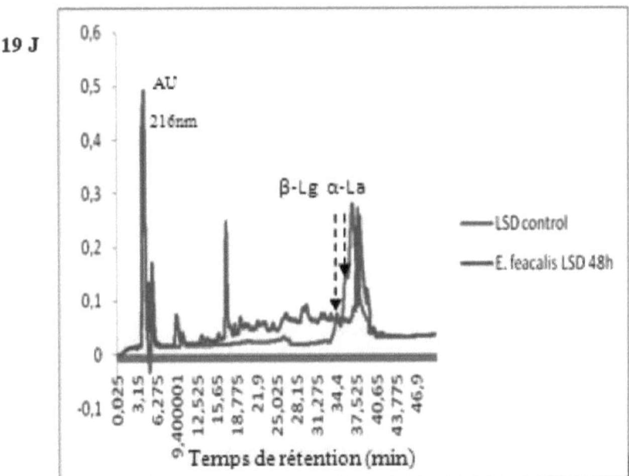

Fig. 19 J HPLC chromatographic analysis of peptide fractions from hydrolysis of whey
denatured by *E. faecalis DAPTO 512* after 48h.
Hydrophobic peptides are eluted at a linear gradient of between 30% and 50% of the mobile

phase, the retention time for these peptides is between 17-36 min.

1.5.3 Effect of pH and temperature on the proteolytic activity of lactic acid bacteria

To determine the effect of temperature and pH on protease activity, strains were cultured on MCA medium (Fira et *al.*, 2001). Collected cells were placed in 100mM phosphate buffer pH 6.6 and pH 5.4 (10 $_{OD600}$, V/V). They were then mixed with sodium caseinates (12mg/ml). The mixtures and the control were incubated for 48 h at two increasing temperatures: 37°C and 40°C. At the end of the incubation period, the supernatant was collected for electrophoretic analysis (fig.20).

The electrophoretic profile of caseinates hydrolysed by the *E. faecalis* strain shows that optimal proteolytic activity is obtained at 37°C and a pH of 6.6.

In contrast, analysis of the electrophoretic profile of the *L. paracasei* strain indicates high proteolytic activity at 40°C at a pH of 6.6. Low molecular weight peptides are produced < 24 KDA and probably correspond to the degradation of α-S2 and β-casein.

For the *E. faecium* strain, our results show total hydrolysis of the three casein fractions at pH 6.6 at 37°C and 40°C. However, the hydrolysis of β-caseins and as2-caseins is partial at acid pH.

Concerning the *E. faecalis* DAPTO 512 strain, our results are in agreement with those of Fira et *al.* (2001), El-Ghaish et *al.* (2010) and Ahmadova et *al.* (2011) who show that the majority of the lactic acid bacteria studied had good proteolytic activity at pH 6.6 and 37°C. It was also shown that optimal conditions for proteolytic enzymes were obtained in an *E. faecalis* strain at neutral pH (El-Ghaish et *al.*, 2010).

For both *L. paracasei* and *E. faecium* strains, our results concur with those of Fira et *al,* (2001) who showed that the optimum temperature for the degradation of β-casein by *L. acidophillus* strain BGRA43 and *L. delbruekii* strain BGPF1 was between 40°C and 45°C. Gardini et *al* (2001) reported that pH is a key factor influencing the activity of amino acid decarboxylases. Koessler et *al* (1928) suggested that the formation of amines by lactic acid bacteria was a physiological mechanism for reducing the acidity of the environment.

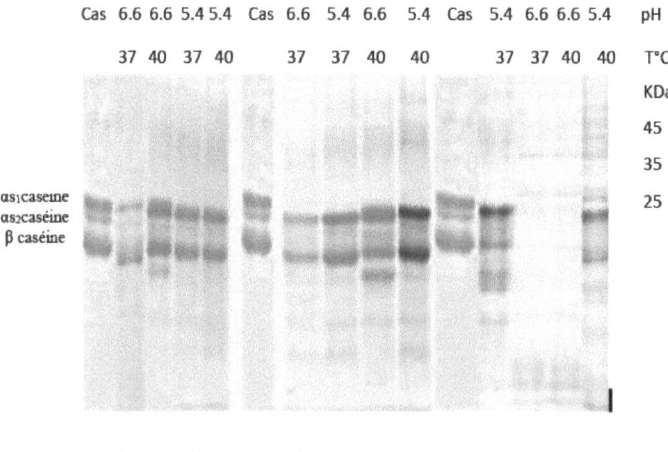

Fig.20 SDS-PAGE analysis (12%) of sodium caseinates at pH 5.4, 6.6 and temperatures 37°C and 40°C.

1.5.4 Effect of inhibitors on proteolytic activity

To determine the nature of the proteases involved in the proteolytic activity of the selected strains. Proteolysis inhibitors were used in this work.

The strains were cultured again on MCA medium (Fira et *al.*, 2001). Collected cells were suspended in phosphate buffer pH 7.2. Various proteolysis inhibitors were used EDTA: Ethylene Diamine Tetra-Acetide to inhibit metallo-proteases, iodoacetic acid to inhibit cysteine proteases, and PMSF to inhibit serine proteases. These inhibitors were added to cell suspensions (10 OD600) at a final concentration of 10 mM, and incubated for 1 h at 37°C before adding the substrate. The inhibitors used were diluted in 10 mM phosphate buffer pH 7.2 to a concentration equivalent to that of the other inhibitors. The samples were centrifuged (10 min at 10,000 rpm), then the cells were mixed with sodium caseinates substrate at 12 mg/ml; 1:1, V/V. The mixtures and the control were incubated for up to 2 nights to assess proteolytic activity against sodium caseinates by SDS-PAGE (12%).

The results show (Fig.21) a significant decrease in proteolytic activity for both *E. faecium* and *E. faecalis* strains in the presence of EDTA and compared with the control without inhibitor (no inhibition). However, the addition of iodoacetic acid and PMSF did not affect the function of the proteases, indicating that the proteases are metalloproteases. This result is in agreement with that of El-Ghaish et *al*, (2010); Ahmadova et *al*, (2011). The proteolytic activity of *L. paracasei* did not change after the addition of iodoacetic acid. However, a slight decrease was observed after the addition of the metallo-protease inhibitor EDTA and a significant decrease in the presence of the serine protease inhibitor PMSF compared to the control (cells without inhibitors). A study by Tsakalidou et *al* (1999) showed that proteases in *L. delbruekii* are strongly inhibited by PMSF but slightly inhibited by EDTA. Based on this same concept, several hypotheses have been proposed concerning the presence of at least two types of proteinases on the cell surface of certain lactobacilli (Stefanitsi et *al.*, 1995; Gilbert et *al.*, 1997; El-Ghaish et *al.*, 2010; Ahmadova et *al.*, 2011).

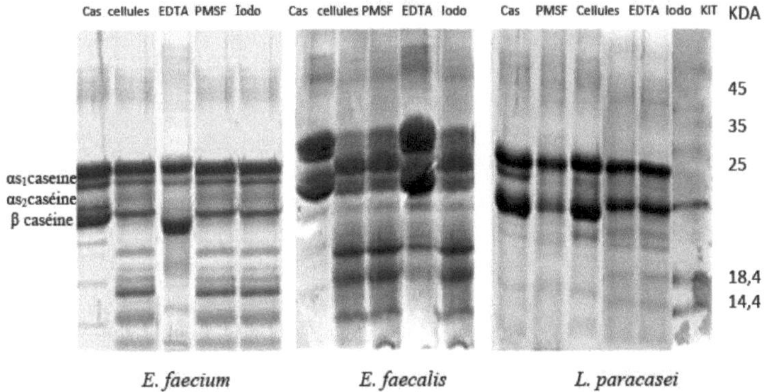

Fig.21 Electrophoresis analysis by SDS-PAGE (12%) of bacterial cells on sodium caseinates of *E. faecium* , *E. faecalis DAPTO 512* and *L. paracasei* in the presence of EDTA, PMSF and Iodoacetic acid, compared to sodium caseinates without cells or sodium caseinates with cells without inhibitors .

1.6 Determination of protein concentration in fermented milk hydrolysates

The results show a decrease in protein concentration in the different fermented milk hydrolysates compared with the initial concentration of milk taken as a control (fig.22). All the strains used degraded the milk proteins to varying degrees, which is probably due to the growth potential and the proteolytic system. This result is in line with those of Kunji et *al*, (1996), who show that lactic bacteria have a complex proteolytic system capable of degrading milk caseins into peptides and free amino acids to ensure their growth in milk. Christensen et *al*, (1999), reported that the growth of lactic bacteria in milk is linked to the proteolytic system, with the release of peptides essentially derived from caseins and that the conversion of these peptides into free amino acids is part of the metabolic activity of the bacteria.

1.7 Determination of the antioxidant activity of hydrophobic peptide fractions from different fermented milk hydrolysates

This test is based on the ability of an antioxidant to stabilise the cationic radical $ABTS^+$ (2,2'-azino-bis-(3-ethyl-benzothiazoline-6-sulfonic acid) (ABTS: Sigma-Aldrich) from blue-green coloration by transforming it into colourless $ABTS^+$, by trapping a proton by the antioxidant. A comparison was made with the ability of Trolox (a water-soluble structural analogue of vitamin E) to capture ABTS+. The decrease in absorbance caused by the antioxidant reflects the free radical's capture capacity.

Fig. 23 shows that the trolox equivalent antioxidant capacity (TEAC) of the different peptide hydrolysates is different. The highest TEAC value, is observed in the *L. paracasei* hydrolysate with a value of (19.2 ± 0.07µM) compared to the milk hydrophobic fractions with (3.75 ± 1.06 µM/g) followed by the TEAC of the hydrolysate peptide fractions (Strp-Blg) (16 ± 0.1) and *L. plantarum-Bifidobacterium longum* (Blg-Lp) (13 ± 0.08 µM), *E. faecalis* (10.3±1.02 µMg) and the TEAC of the *E. faecium* hydrolysate fractions with (8 ± 0.1 µMg).

Our results are in line with those of numerous researchers, as it has been shown that after degradation of milk protein sequences by microbial proteases or plant enzymes, bioactive peptides can be released (Hayes et *al.*, 2007; Korhonen & Pihlanto, 2006). In 2006 Pihlanto mentioned that certain peptides derived from milk proteins were considered to be a new class of antioxidants.

The TEAC of the hydrophobic fractions of *L. paracasei* fermented milk hydrolysates increased very significantly compared to that of cow's milk. This result is in agreement with that found by Moslehishad et *al,* (2013), she showed that peptides from cow's milk and camel's milk fermentation significantly increased antioxidant activity following hydrolysis of α-S1 casein and β-casein. It was also reported by Hernández-Ledesma et *al*, (2013) that milk protein hydrolysates modified mucus dynamism via secretions and expression of numerous caliciform cells. Authors have reported that Kefir, yoghurts and fermented milks are commercialised dairy products that can obtain bioactive peptides following proteolysis and microbial fermentation (Muri Urista et *al.*, 2011) and it would appear that *Streptococcus thermophillus* is a good candidate for producing fermented milks with functional properties (Kunji et *al.*, 1996; Christensen et *al.*, 1999; Savijoki et *al.*, 2006, Hafeez et *al.*, 2013).

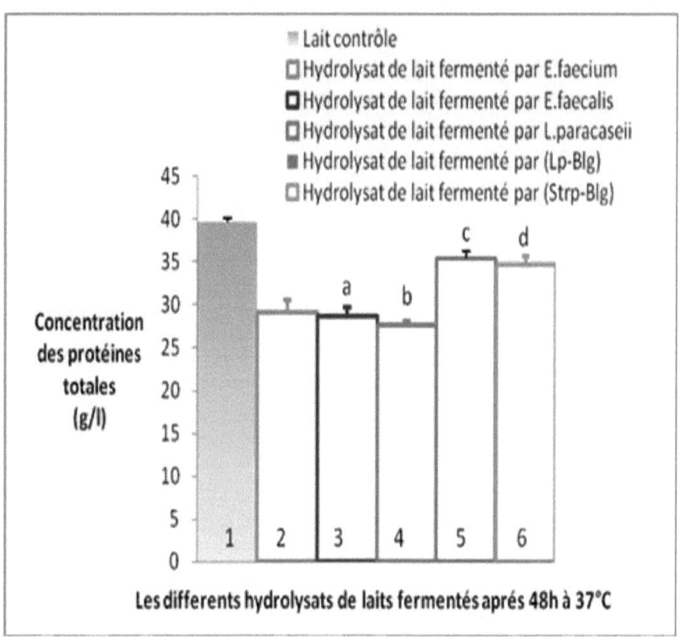

Fig. 22 Protein concentration in (g/l) of fermented milk hydrolysates after 48 h of fermentation at 37°C.

1. Control milk; 2.hydrolysate of milk fermented by *E. faecium;* 3.hydrolysate of milk fermented by *E. faecalis DAPTO 512*; 4.hydrolysate of milk fermented by *L. paracasei*; 5. Milk hydrolysate fermented with *L. plantarum-Bifidobacterium longum* (Lp-Blg); 6. *Streptococcus thermophillus-Bifidobacterium longum* (Strp-Blg) fermented milk hydrolysate.

The (*p*) values represent comparisons of the means X±SE (n=3), of the protein concentrations of the hydrolysates with the control milk.

[*]*p E. faecium* LF *hydrolysate p<0.001.*[a] *p E. faecalis* DAPTO 512 hydrolysate *p<0.001.*[b] *p L. paracasei* hydrolysate *p<0.001.*[c] p hydrolysate (Lp-Blg) *p<0.01.*[d] p hydrolysate (Strp-Lp) *p<0.01.*

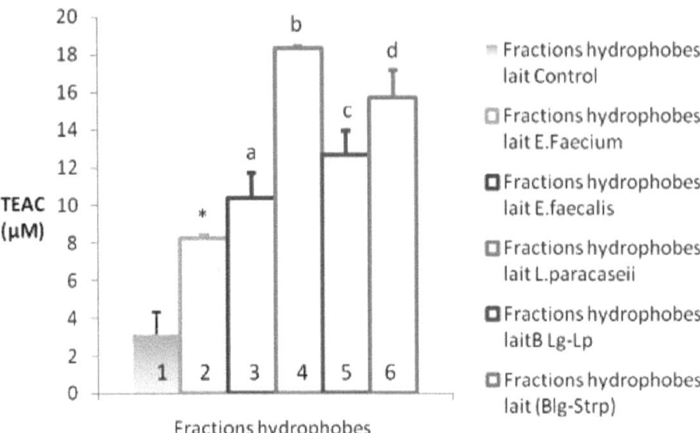

Hydrophobic milk fractions Control
☐ Hydrophobic milk fractions E.Faecium
☐ Hydrophobic fractions EJaecalis milk
☐ Hydrophobic milk fractions L.paracaseii
☐ Hydrophobic fractions laitB Lg-Lp
☐ Milk hydrophobic fractions (Blg-Strp)

Fig.23 Antioxidant activity of hydrophobic peptide fractions of different milks fermented by lactic acid bacteria after 48h incubation at 37°C.

Values are presented as X±SE (n=3), [*] p represents the comparison of trolox antioxidant equivalent capacity between sample 2 and sample 1 *p<0.01*.[a] p represents the comparison of trolox antioxidant equivalent capacity between sample 3 and sample 1 *p<0.001*. [b]*p* represents the comparison of the trolox equivalent antioxidant capacity between sample 4 and sample 1 *p<0.000*.[c] p represents the comparison of the trolox equivalent antioxidant capacity between sample 5 and sample 1 *p<0.01*.[d] p represents the comparison of the trolox equivalent antioxidant capacity between sample 6 and sample 1.

1.8 Inhibitory action of lactic acid bacteria against indicator and/or pathogenic bacteria

- on *Escherichia coli*

The antimicrobial activity of the above bacteria was determined by the well diffusion method according to Schillinger and Lucke (1989). The test was carried out using bacterial cultures and their supernatants by centrifugation (10000 t, 15min, 4°C), the pH of the supernatant was adjusted to 6.5 with 1N NAOH.

One positive result was observed for the *L. paracasei* strain, the presence of halos with a diameter of 4mm was considered positive (Table 12). No results were observed for *E. faecium, E. faecalis*. This result is in agreement with that found by Sisto et *al.* (2012). He reported that a strain of *lactobacillus paracasei* had the ability to inhibit the growth of *Escherichia coli* and *Clostridium* spp. however it does not agree with the studies carried out by De Vuyst and Vandamme, (1994), Batdorj et *al.*(2006), Ben Belgacem et *al.,* (2010), Giraffa, (2003), Yammamto et *al.*,(2003). These studies have shown that Enterococci produce bacteriocins called enterocins, defined as small peptides with inhibitory activity against Gram-positive bacteria and the ability to damage pathogenic bacteria.

To determine the nature of the inhibiting agent, the samples were treated with catalase and proteinase K in phosphate buffer (0.1M, pH 7.0), acidic conditions (pH 4.5) and neutralised. Two hundred μl of the cell culture or its supernatant was mixed with these enzymes at a final concentration of (0.1 mg/ml). After 2 h of incubation the reaction was stopped after heating for 3 min. The reaction was tested a second time with *E.Coli* (fig.24).

The result shows the presence of halos around well 1: which represents the growth of *L. paracasei* under acidic conditions, and around well 3 where the supernatant under acidic conditions was deposited. These zones disappeared under neutralised conditions. The appearance of these clear zones does not confirm that these inhibitions are due to a bacteriocin; they may be due to the acidification of the medium. Desmazeaud (1983) dealt with this criterion. This is why the bacterial culture was neutralised to eliminate the effect of acidity on the medium. After treatment with proteinase K, the proteolytic enzyme, these two halos disappeared, so we can say that the active agent is proteinaceous. This result is in line with what was reported by two researchers, Callewaert and de Vuyst, (2000) and Aslim et *al*, (2005). This indicates that this substance is probably not a bacteriocin as it has been inhibited by one of the proteolytic enzymes, and the loss of antimicrobial activity after treatment with enzymes indicates the sensitivity of the active compounds secreted by the cultured strain. Consequently, many authors claim that sensitivity to proteolytic enzymes is the main criterion in their characterisation (Icon and Gökalp, 2000; El Shafei et *al.*, 2000; Cintas et *al.*, 2001).

The sample was also treated with the catalase enzyme, and no zone of inhibition was observed. This confirms the sensitivity of the active agent to hydrogen peroxide. Juillard et *al* (1987) reported that hydrogen peroxide has long been recognised as a major agent of antimicrobial activity.

Table 12: Inhibitory activity of bacterial strains against *Escherichia coli* (E. Coli) ATCC 23355

Bacterial strains		*E.coli*
LParacasei	Culture	4mm
	Supernatural	+/-
E. feacium	Culture	-
	Supernatural	-
Efaecalis	Culture	-
	Supernatural	-

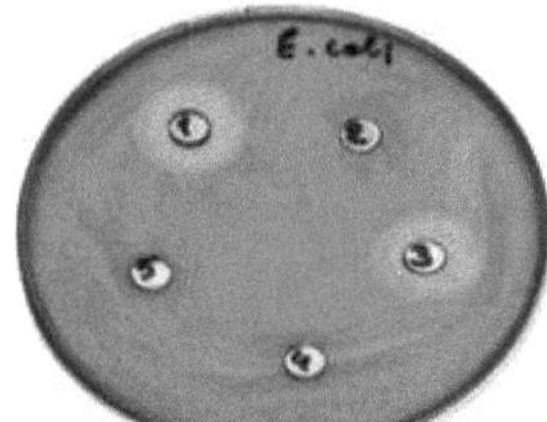

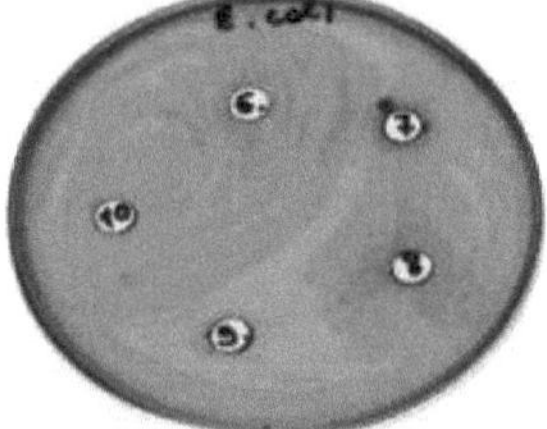

Fig.24 Inhibitory activity of the *L. paracasei* strain *against E.coli*.

1- Bacterial culture for 24 hours under acidic conditions pH 4.5;
2- Bacterial culture for 24 hours under acidic conditions, then neutralised to pH 6.5;
3- Supernatant of 24h bacterial culture in acidic conditions pH 4.5;
4- Supernatant of 24h bacterial culture under acidic conditions and neutralised to pH 6.5;

5- Neutralised supernatant + proteinase K;

6- Neutralised supernatant + catalase;

7- Supernatant neutralised with a small crystal of catalase next to the well;

8- 9-10- Negative controls (liquid SRM without culture, with catalase and with proteinase K).

1.9 Growth of lactic acid bacteria in MRS medium at different pH and bile concentrations

The ability of proteolytic lactic acid bacteria to grow at different pH levels and in different concentrations of bile salts was tested and enumerated to assess the bacterial concentration in CFU/ml.

1.9.1 Effect of pH on bacterial growth

After one night's incubation at 37°C, weak to absent bacterial growth was observed for all strains tested at pH 3.0 *(< 3 log cfu/ml)*. At pH 6.0, average growth for both *E. faecium* and *E. faecalis* strains ranged from 5 log cfu/ml to 6 log cfu/ml. At pH 7.0, 9.0, 11.0 growth was maximal (8 log cfu/ml and 9 log cfu/ml) and decreased at pH 13.0 (between 7 log cfu/ml and 6 log cfu/ml). No growth of either strain was observed at pH3.0 and pH4.0. These results are in line with those of Giraffa (2003) concerning the optimal pH for Enterococci.

For Lactobacilli, between pH 3.0 and 4.0 average growth was observed (between 5 log and 6 log cfu/ml). Above this pH, the two lactobacilli showed different profiles. *L. paracasei* showed good growth at pH 5.0, 6.0, 7.0, 9.0 (8 log and 9 log cfu/ml) and at pH 11.0 and 13.0 a significant decrease was observed. Concerning the growth of *L. plantarum,* at pH 4.0, 5.0, 6.0, 7.0 the concentration was between (8 log and 9 log cfu/ml), and it saw a decrease from pH 9.0 (±5 log cfu/ml) to reach (< 4 log cfu/ml) at pH 11.0 and 13.0. Similarly, in *L. paracasei* a significant decrease was observed at pH 11.0 and 13.0 (±5 log cfu/ml). These results were similar to those of Sisto et *al,* (2012), who compared a strain of *L. paracasei* with several strains such as *L. plantarum* and *L. fementum* and found, after *in-vitro* tests, that *L. paracasei* was resistant to gastric juice and bile salts. Conway et *al,* (1987); Du Toit et *al,* (1998); Jacobsen et *al,* (1999); Dunne et *al,* (2001); Maragkoudakis et *al,* (2006); Xanthopoulos et *al,* (2000) have shown that lactobacilli of food or animal origin are able to retain their viability at pH levels between 2.5 and 4.0.

The *Streptococcus thermophillus* strain grew well between pH 5.0 and 11.0 (±8 log cfu/ml). At pH 13.0 growth slowed (6 log cfu/ml). For the *Bifidobacterium longum* (Blg) strain, we observed very weak to absent growth between pH 3.0, 4.0 and 5.0 (≤ 10 log to 3 log cfu/ml), with good growth at pH 7.0 and 9.0 (8 log and 9 log cfu/ml). These results concur with those of Zacarias et *al* (2011), who showed that culture of a *Bifidobacterium animalis subsp. lactis* strain isolated from human breast milk could rapidly reach its stationary phase at pH 6.5 (Table 13).

1.9.2 Effect of bile on bacterial growth

All the *E. faecalis* and *E. faecium* strains grew well in the presence of different concentrations of bile ranging from 0.2% to 3% (Table 13). Good growth was observed for *L. plantarum* and *L. paracasei* as well as *Streptococcus thermophillus* at 0.2% and 0.3% bile (±8 log cfu/ml). A highly significant decrease was observed at 0.5% bile, with concentrations of (<4 log cfu/ml; <6 log cfu/ml; <4 log cfu/ml) respectively. While *Bifidobacterium longum* survived only 0.2% bile. In another study by Strahinic et *al,* (2012), it was shown that the viability of *L. helveticus* was reduced by three log units from 0.3% to 0.6% bile. While Argyri et *al.* (2013) tested lactobacilli in high concentrations of bile, the majority of strains managed to survive after 3 hours.

1.10 Assessment of antibiotic resistance

Because of the risks of transmitting antibiotic resistance genes in foodstuffs to other bacteria. It is necessary to assess not only their technological characteristics, but also the safety of the strains before application in food. In this study, the susceptibility of different bacterial strains to antibiotics was investigated.

Our results (Table 14) showed that *E. faecalis* DAPTO 512 and *E. faecium* are susceptible to ampicillin (Minimum Inhibitory Concentration (MIC) < 8Lig/ml), penicillin (MIC < 4µgZml) gentamycin (MIC <500 Lig/ml), tetracycline (MIC < 4µgZml), vancomycin (MIC<4µg/ml), ciprofloxacin (MIC <4µg^l) and chloranphenicol (MIC=8µg^l).

Table 13: Growth of the different strains used under different bile and pH conditions

Lactic acid bacteria	Bile concentration % (%)								pH							
	0	0,2	0,3	0,5	0,6	1,0	2,0	3,0	3,0	4,0	5,0	6,0	7,0	9,0	11,0	13,0
E. faecium	++	++	++	++	++	++	++	++	+/-	+/-	-	+	++	++	++	+
E. f(tecalis DAPTO512	++	++	++	++	++	++	++	++	+/-	+/-	-	+	++	++	++	+
L. paracasei	++	++	++	+	-	-	-	-	+/-	+	++	++	++	++	-	-
Bifidobacterium Iongum	++	+	-	-	-	-	-	-	-	-	-	+	++	++	+	-
Streptococcus thermophillus	++	++	++	+	-	-	-	-	-	+/-	++	++	++	++	++	-
L. plantarum	++	++	++	-	-	-	-	-	+/-	+	++	++	++	+	-	-

Table 14: Minimum inhibitory concentrations (MICs) of antibiotics obtained for the different strains

Antibiotics	Susceptibility to antibiotics MIC (µgZml)[1]					
	L. paracasei	*E. faecalis*	*Efaecium*	*Lplantarum (Lp)*	*Streptococcus thermophillus* (StrP)	*Bifidobacterium Longum (Blg)*
Penicillin (P)	<4	<4	<4	>4	<4	<4
Ampicillin (AM)	<8	<8	<8	<8	<8	<8
Chloramphenicol (C)	= 8	= 8	= 8	= 8	= 8	= 8
Ciprofloxacin (CIP)	<32	<4	<4	<64	<32	<32
Gentamycin (GEN500)	<500	<500	<500	<500	<500	<500
Vancomycin (VA)	<16	<4	<4	<4	<4	<4
Kanamycin (K)	ND	ND	ND	ND	ND	=8
Tetracycline(TE)	<2	<4	<4	<4	<4	<4

This result is in line with that of Ben Omar et *al*, (2004) and Veljovic et *al*,(2009), where they observed that enterococci in food were sensitive to ß-Lactamins.

The sensitivity of Lactobacilli to antibiotics was established according to a sensitivity threshold defined as follows: Ampicillin- 4 µg/ml, Chloramphinicol- 16 µg/ml, Penicillin- 4µg/ml, Tetracycline -16 µgZml, Ciprofloxacin - 4µgZml, Vancomycin -64 µgZml. They are all resistant to gentamycin (no sensitivity up to 500 µgZml). *L. plantarum* is resistant to ampicillin while *L. paracasei* is susceptible (MIC<2µgZml) and sensitive to other antibiotics (MIC is lower than the sensitisation threshold).

Results similar to this work have shown that lactobacilli are usually sensitive to ampicillin and chloramphinicol (Ammor et *al.*, 2007; Katla et *al.*, 2001). While Mikelsaar et *al*, (2004) showed that 100% of the *L. plantarum* strains tested were resistant to vancomycin and 71% of them resistant to ciprofloxacin, Kmet et *al*, (1989) found *that L. acidophillus* and *L. reuteri* were resistant to most of the antibiotics tested. The use of lactic acid bacteria in the manufacture of dairy products has increased significantly. In recent years, the issue of antibiotic resistance has been raised (Ammor et *al.*, 2007). The excessive use of antibiotics in humans and animals has created resistance that can be transmitted to food. The evolution of microbial resistance to antibiotics in food has been reported (Zhao et *al.*, 2006; Rahimi et *al.*, 2010; Gousia et *al.*, 2011). Recently, resistance to Vancomycin has evolved considerably and is the primary cause of nosocomial infections. According to the literature, to date very few isolates are resistant to it (Barbosa et *al.,* 2001). *Streptococcus thermophillus* and *bifidobacterium longum* are sensitive to all antibiotics except kanamycin and gentamycin. Masco et *al*, (2006) and Ammor et *al*,(2007) have shown that bifidobacteria are generally susceptible to vancomycin and β-lactam antibiotics and resistant to kanamycin and gentamycin. Zhou et *al*, (2011) showed that a strain of *Streptococcus thermophillus* isolated from Chinese yoghurt had potentially transferable resistant genes during their passage through the food chain.

2. Evaluation of the therapeutic/preventive effect of cow's milk hydrolysates fermented by selected lactic acid bacteria

2.1 Assessment of the preventive effect

2.1.1 Measurement of the preventive effect of orally administered hydrolysates on the intestinal mucosa of Balb/c mice following intraperitoneal sensitisation to ß-Lg

2.1.1.1 Anti-ß-Lg IgG assay

At the end of the experiment, serum IgG anti-ßLg antibodies were produced to a highly significant degree *(p<0.001)* in the positive control group (1/1000000^{éme}), whereas they were absent in the negative control group. For the other groups of mice that received fermented milk hydrolysates for 15 days before sensitisation, a low production of anti-ß-lg IgG was observed to varying degrees (fig.25).

The lowest IgG titre was detected in the LF *L. paracasei* group with a very significant titre of (1/100^{éme}), followed by the LF *E. faecium* group (1/1000^{éme}), and the LF *E. faecalis* group with a titre of (1/1500^{éme} *), with* the two LF (Lp-Blg) and LF (Strp-Blg) groups ranking last with (1/10000).^{éme}

2.1.1.2 Anti-ß-Lg IgE assay

After 35 days of immunisation (D35) the serum anti-ß-lg IgE titre increased highly significantly *(p<0.001)* in the positive control group, whereas it was undetectable in the negative control group. For the other groups of mice that received fermented milk hydrolysates for 15 days, a decrease in the production of anti-ß-lg IgE was observed to varying degrees (fig.26).

The most significantly lower titre was observed in the LF *L. paracasei* group compared to the positive control, followed by the LF *E. faecium* and LF *E. faecalis* groups respectively *p<0.01* and the LF (Blg-Lp), *p<0.01* and LF (Strp-Blg) *p<0.01* groups.

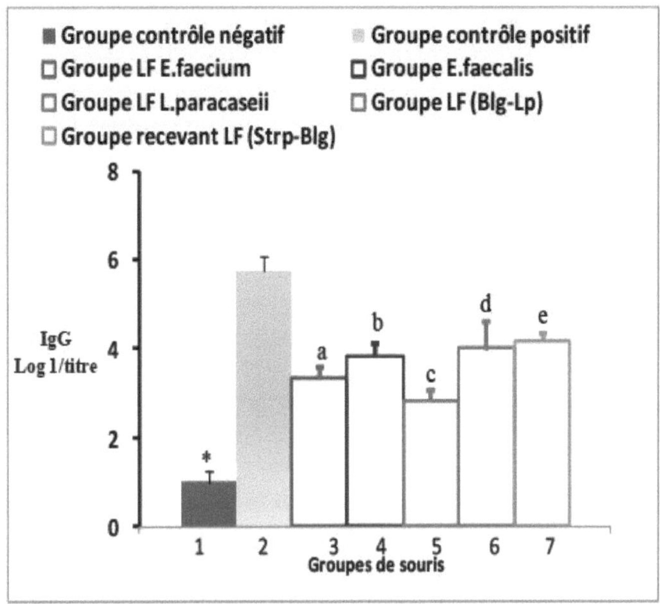

Fig.25 Serum anti β-Lg IgG titers measured at D35 in mice given hydrolysates orally and then sensitised intraperitoneally to β-Lg (n=6).

Group 1negative **control**; Group 2positive **control**; Group **3** receiving oral milk hydrolysates obtained after fermentation by *E. faecium* and then sensitised intraperitoneally to β-Lg; Group **4** receiving milk hydrolysates fermented by E. faecalis and then sensitised intraperitoneally to β-Lg; Group **5** receiving milk hydrolysates fermented by *L. paracasei* and then sensitised intraperitoneally to β-Lg; Group 6 receiving milk hydrolysates fermented by the combination (Blg-lp) and then sensitised intraperitoneally to β-Lg. *paracasei* then sensitised intraperitoneally to β-Lg; Group **6** receiving hydrolysates of milk fermented with the combination (Blg-lp) then sensitised intraperitoneally to β-Lg; Group **7** receiving hydrolysates of milk fermented with the combination (Strp-Blg) then sensitised intraperitoneally to β-Lg.

Values are presented as X±SE (n=6),

The *(p)* values represent the comparison of the mean anti β-Lg IgG titres of the different groups with the positive control group,[*] p represents the comparison between group **1** and group **2** *p<0.0001.*[a] p represents the comparison between group 3 and group 2 *p<0.01.*[b]*p* represents the comparison between group 4 and group 2 *p<0.01.*[c] p represents the comparison between group 5 and group 2 *p<0.001.*[d] p represents the comparison between group 6 and group 2 *p<0.01.*[e] p represents the comparison between group 7 and group 2 *p<0.01.*

We note that levels of anti-β-Lg IgG production are undetectable at D0.

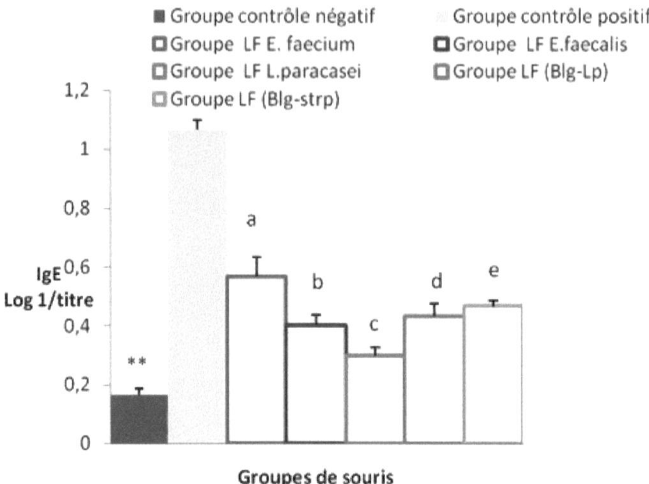

Fig.26 Serum anti ß-Lg IgE titers measured at D 35 in mice given hydrolysates orally and then sensitised intraperitoneally to β -Lg (n=6).

Group **1** negative control; Group **2** positive control; Group **3** receiving oral hydrolysates of milk fermented by *E. faecium* then sensitised intraperitoneally to ß-Lg; Group **4** receiving hydrolysates of milk fermented by E. faecalis then sensitised intraperitoneally to ß-Lg; Group **5** receiving hydrolysates of milk fermented by *L. paracasei* then sensitised intraperitoneally to ß-Lg; Group **6** receiving hydrolysates of milk fermented with the (Blg-strp) combination then sensitised intraperitoneally to ß-Lg; Group **7** receiving hydrolysates of milk fermented with the (Blg-Lp) combination then sensitised intraperitoneally to ß-Lg.

Values are presented as X±SE (n=6), (***p***) **values** represent the comparison of the mean anti ß-Lg IgE titres of the different groups with the positive control group,[**] p represents the comparison between group 1 and group 2 *p<0.001.*[a] p represents the comparison between group 3 and group 2 *p<0.01.*[b]*p* represents the comparison between group 4 and group 2 *p<0.01.*[c] p represents the comparison between group 5 and group 2 *p<0.01.*[d] p represents the comparison between group 6 and group 2 *p<0.01.*[e] p represents the comparison between group 7 and group 2 *p<0.01.*

It should be noted that levels of anti-B-Lg IgE production were undetectable at D0.

2.1.1.3 *In vitro* Ussing chamber challenge test

The aim of this part of the work is to assess the local anaphylactic response in mice sensitised with the allergen and to measure the level of immunomodulation induced in animals given the different hydrolysates.

2.1.1.4 Effect of ß-Lg on the short circuit current (Isc) measured in the Ussing chamber on jejunal fragments from mice given fermented milk hydrolysates and then sensitised intraperitoneally to ß-Lg.

After mounting a jejunal fragment of intestine in an Ussing chamber and after stimulation with the ß-Lg allergen at a concentration of 60 µg/ml in the serous compartment, we measured the short-circuit current (Isc). Our results show that stimulation with the allergen causes a significant increase in Isc (Δ Isc = 24µA/cm^2) in the positive control group (Fig.27). In the negative control group, Isc is (Δ Isc = 2 µA7cm^2).

In general, the response to allergen stimulation was significantly lower in the groups that received the different fermented milks than in the positive control. A decrease in the short circuit current (Isc) was observed in the LF *L. paracasei* group; stimulation with the allergen resulted in a slight increase in Isc (Δ Isc = 8µAZcm^2) but it remained significantly lower than that of the positive control group (ΔIsc=24 µA7cm^2) *p<0.0001*(fig.27).

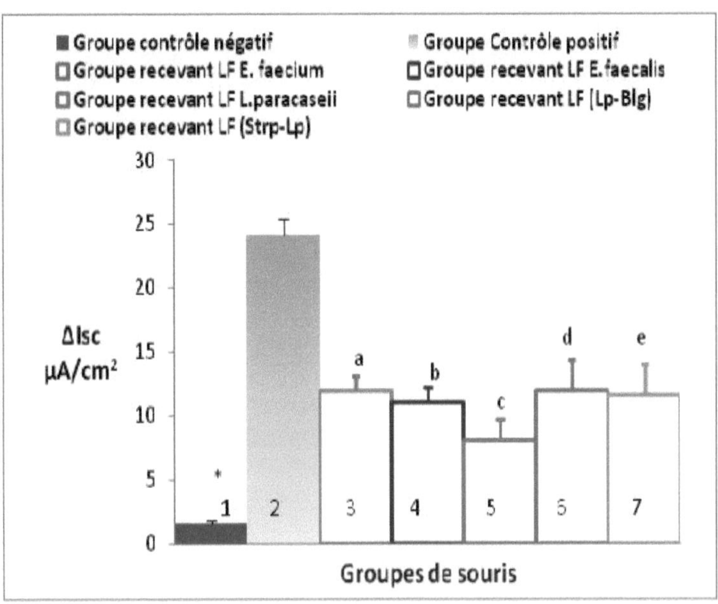

Fig.27 Effect of ß-Lg on the short circuit current (Isc) at D 35 measured in the Ussing chamber on jejunal fragments from mice receiving fermented milk hydrolysates and then sensitised intraperitoneally to β-Lg (n=6).

Group 1 negative control; Group 2 positive control; Group 3 receiving oral   milk hydrolysates   obtained after fermentation   by   *E.*   faeciuissensitisedbyvariety intraperitoneal   to   ß-Lg; Group 4 receiving   LF   E.
     faecalisfissensitisedbyvessel
intraperitoneally with ß-Lg; Group 5 receiving LF *L. paracasei* followed by intraperitoneal sensitization   with   ß-Lg;   Group 6 receiving LF (Blg-lp)   and   then intraperitoneally sensitised to ß-Lg.
intraperitoneal   to   ß-Lg; Group 7 receiving   LF   (Strp-Blg)
     potent   sensitisation via
intraperitoneal to ß-Lg.
X±SE **values** (n=6) represent the comparison of Isc values of the different groups sensitised intraperitoneally with ß-Lg and then with the positive control. *p represents the comparison of Isc between group 1 and group 2, *p<0.0001.*[a] *p* represents the comparison of Isc between group 3 and group 2, p<0.*001.*[b] p represents the comparison of Isc between group 4 and group 2, *p<0.001.* [c]*p* represents the comparison of Isc between group 5 and group 2, *p<0.0001.*[d] p represents the comparison of Isc between group 6 and group 2, *p<0.001.*[e] p represents the

comparison of Isc between group 7 and group 2, *p<0.001.*

2.1.1.5 Measurement of the conductance (G) of mouse jejunal fragments mounted in Ussing chambers in response to ß-Lg stimulation

Table 15 shows the variations in the conductance of tissues from animals in contact with the ß-Lg allergen. The conductance of the tissues studied after stimulation by the sensitising antigen increased significantly in the positive control group, rising from 32.68 ± 1.09 mmho/cm^2 to 43.25 ± 1.31 mmho/cm^2 (*p<0.001*). Tissue conductance in the groups of mice given fermented milk hydrolysates was significantly lower than in the positive control group. The conductance of the negative control group of mice remained unchanged.

2.1.1.6 Effect of sensitising protein (ß-Lg) on short circuit current after deposition of Furosemide.

In order to determine in our model of anaphylaxis the nature of the mechanisms involved in the increase or stability of the secretory activity of the intestinal epithelium in the presence of a sensitising antigen, a diuretic, furosemide, was used. This agent is a specific inhibitor of the Cl/Na/K cotransport system located on the basolateral membrane. Furosemide depletes the cell of Cl$^-$ and therefore reduces the secretion of Cl$^-$

After mounting the jejunal fragments of the mice in the Ussing chamber and stabilising the electrophysiological parameters. Furosemide at a concentration of 5.10^{-2} M was applied to the serosal side. The tissues were then stimulated with the sensitising antigen.

Our results show that the tissues of the positive control group, when previously incubated with furosemide, the sensitising antigen does not stimulate the short circuit current. The same results were observed in the groups of mice that received the hydrolysates (fig. 28). This indicates that the secretion observed under the experimental conditions is indeed a chlorine secretion.

To check the viability of the tissues at the end of the experiment, we tested the effect of glucose at a final concentration of 50mM on the mucous and serous compartments of the tissues. Our results show that glucose stimulates the short circuit current corresponding to the sodium current stimulated by the sodium/glucose symport mechanism.

Table 15: The conductance (G) of jejunal fragments from Mice receiving fermented milk hydrolysates and then sensitised intraperitoneally to β-Lg mounted in Ussing chambers in response to β-Lg stimulation (60µg/ml).

	G (mmho/cm^2		
	Before deposit	P-Lg (60 µg7ml)	After deposit
Negative control group	$26,40 \pm 1,02$	$27,10 \pm 2,1$*	
Positive control group	$32,68 \pm 1,09$		$43,25 \pm 1,31$
Group receiving LF *Efaecium*	$27,56 \pm 1,59$		$31,05 \pm 1,96$ [a]
Group receiving LF *Efaecalis DAPTO 512*	$28,27 \pm 02,2$	$30,12 \pm 2,06$ [b]	
Group receiving *L.paracasei*	$24,07 \pm 2,13$	$28,43 \pm 1,5$ [c]	
Group receiving LF (Blg-Lp)	$24,08 \pm 1,2$	$29,32 \pm 2,29$[d]	
Group receiving LF (Strp-Blg)	$25,23 \pm 2,72$	$30,09 \pm 2,64$[e] [e]	

Values are presented as X±SE (n=6),

The (*p*) values represent the comparison of conductance (G) at D 35 of the groups receiving hydrolysates and then sensitised intraperitoneally to β-Lg with the positive control group.

*p represents the comparison between the negative control group and the positive control group $p<0001$;

[a]p represents the comparison between the group receiving LF *E. faecium* and the positive control group $p<0.01$;

[b]p represents the comparison between the group receiving LF *E. faecalis* and the positive control group $p<0.01$;

[c]p represents the comparison between the group receiving LF *L. paracasei* and the positive control group $p<0.01$;

[d]p represents the comparison between the group receiving LF *(Blg-Lp)* and the positive control group;

[e]p represents the comparison between the group receiving LF (Strp-Blg) and the positive control group.

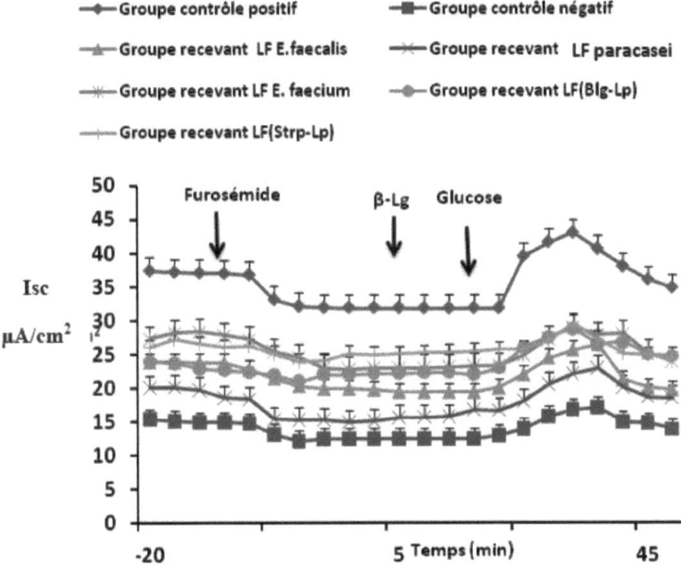

Fig.28 Effect of sensitising protein (ß-Lg) on short circuit current after deposition of Furosemide measured in a Ussing chamber on jejunal fragments of mice receiving hydrolysates of milk fermented by lactic acid bacteria and then sensitised intraperitoneally to ß-Lg (n=6).

2.1.1.7 Non-specific effect of ovalbumin on the short-circuit current measured in the Ussing chamber

In order to show that the increase in Isc of sensitised mouse fragments is specific to the sensitising protein, ovalbumin was tested in the serous compartment (60µg/ml) under the same experimental conditions.

Throughout the incubation period with ovalbumin (10 min experiment), no variation in conductance was recorded (fig.29).

2.1.2 Measurement of the preventive effect of hydrolysates on the intestinal mucosa of Balb/c mice after oral sensitisation to cow's milk

2.1.2.1 Anti-ß-Lg IgG assay

The results show (fig.30) that the positive control group produced a highly significant amount of IgG (1/1000000^{éme}) whereas it was absent in the negative control. Serum anti-ß-Lg IgG titres were reduced in all groups of mice fed fermented milk hydrolysates. The most significantly lower titre was obtained in the LF *paracasei* group (1/100^{éme}) *p<001,* followed by the other groups *p<0.01.*

2.1.2.2 Anti-ß-Lg IgE assay

Fig. 31 shows that the positive control group produced highly significant levels of anti-β-Lg serum IgE antibodies *(*p<0.001), whereas these were absent in the negative control group. The serum IgE titre against ß-Lg was lower in all groups receiving milk hydrolysates compared with the positive control *(p<0.01).*

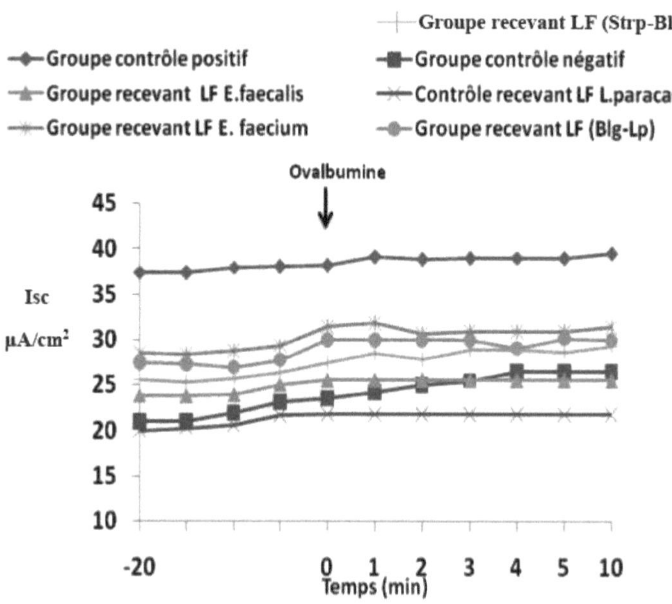

Fig.29 Non-specific effect of ovalbumin on the short-circuit current measured in the Ussing chamber on jejunal fragments from mice given fermented milk hydrolysates and then sensitised intraperitoneally to β-Lg.

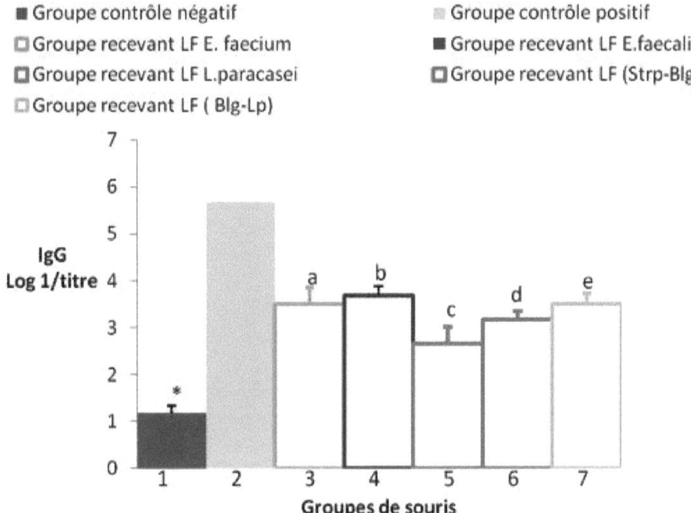

Fig. 30 Serum anti ß-Lg IgG titers in mice at D 35 receiving these hydrolysates orally and then orally sensitised to cow's milk

Group **1** negative control; Group **2** positive control; Group **3** receiving a hydrolysate of milk fermented by *E. faecium* orally then orally sensitised to cow's milk; Group **4** sensitised receiving a hydrolysate of milk fermented by *E. faecalis* DAPTO 512 then orally sensitised to cow's milk; Group **5** sensitised receiving a hydrolysate of milk fermented by *L. paracasei* fermented milk hydrolysate and then orally sensitised to cow's milk; Group **6** sensitised receiving an oral (Blg-Lp) fermented milk hydrolysate and then orally sensitised to cow's milk; Group **7** sensitised receiving an oral (Strp-Blg) fermented milk hydrolysate and then orally sensitised to cow's milk.

Values are presented as X±SE (n=6), **p-values** represent the comparison of serum anti ß-Lg IgG titres in groups of mice receiving hydrolysates of fermented milks and then orally sensitised to cow's milk with the positive control group.[*] p represents the comparison between group 1 and group 2 *p<0.0001.* [a]*p* represents the comparison between group 3 and group 2 *p<0.01.*[b] *p* represents the comparison between group 4 and group 2 *p<0.01.*[c] p represents the comparison between group 5 and group 2 *p<0.001.*[d] p represents the comparison between group 6 and group 2 *p<0.01.*[e] p represents the comparison between group 7 and group 2 *p<0.01.*

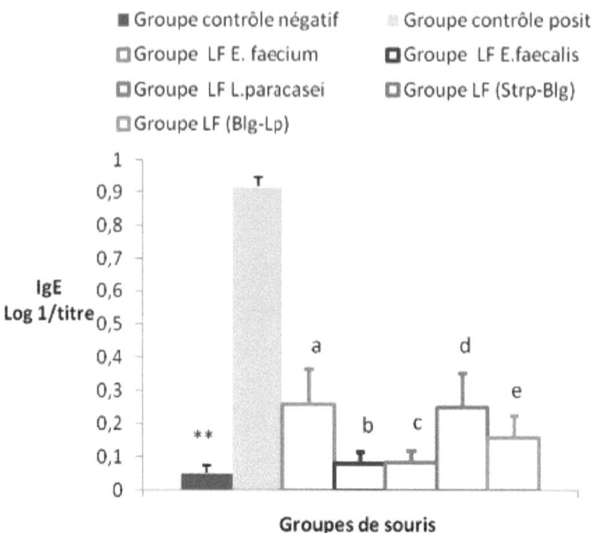

Fig.31 Serum anti ß-Lg IgE titers in mice at D 35 receiving hydrolysates orally and then orally sensitised to cow's milk.

Group **1** negative control; Group **2** positive control; Group **3** receiving oral *E. faecium* fermented milk hydrolysate followed by oral sensitisation to cow's milk; Group **4** receiving oral *E. faecalis* DAPTO 512 fermented milk hydrolysate followed by oral sensitisation to cow's milk; Group **5** receiving oral *L. paracasei* fermented milk hydrolysate followed by oral sensitisation to cow's milk; Group **6** receiving oral (Blg-strp) fermented milk hydrolysate followed by oral sensitisation to cow's milk. *paracasei* fermented milk hydrolysate and then orally sensitised to cow's milk; Group **6** receiving an oral (Blg-strp) fermented milk hydrolysate and then orally sensitised to cow's milk; Group **7** receiving an oral (Blg-Lp) fermented milk hydrolysate and then orally sensitised to cow's milk.

Values are presented as X±SE (n=6). **Values (*p*)** represent the comparison of serum IgE anti ß-Lg titres at D 35 in groups of mice receiving hydrolysates of fermented milk and then orally sensitised to cow's milk with the positive control group.[**] p represents the comparison between group 1 and group 2 *p<0.001*[a] p represents the comparison between group 3 and group 2 *p<0.01.*[b]*p* represents the comparison between group 4 and group 2 *p<0.01.*[c] p represents the comparison between group 5 and group 2 *p<0.01.*[d] p represents the comparison between group 6 and group 2 *p<0.01.*[e] p represents the comparison between group 7 and group 2 *p<0.01.*

It should be noted that levels of anti-B-Lg IgE production were undetectable at D0.

2.2 Evaluation of the therapeutic effect of fermented milk hydrolysates after intraperitoneal sensitisation to ß-Lg

2.2.1 Anti-ß-Lg IgG assay

The results (fig. 32) show that in the positive control group anti-ß-Lg serum IgG is produced in a highly significant manner (1/100000[éme]) (*p<0.0001*).

In general, we observed a weak response in serum IgG production in all the groups receiving the hydrolysates. The lowest level was observed in the group that received LF *L. paracasei* milk hydrolysates with a highly significant titre of (1/100^{éme}) *(p<0.0001)* followed by the group that received LF *E. faecium hydrolysate* (1/1000^{éme}) and LF *E. faecalis* (1/1000^{éme}) *(p<0.01)*, to finish with the two groups that received LF (Blg-Lp) (1/10000^{éme}) *(p<0.01)* and (Strp-Blg) (1/10000^{éme}) *(p<0.01)*.

2.2.2 Anti-ß-Lg IgE assay

Serum anti-ßLg IgE titres in the positive control group were produced to a high degree of significance *(p<0.001)* (fig.33). A decrease was observed in the animals that received the hydrolysates. The lowest titre was observed in the LF *L. paracasei* group *(p<0.01)* compared with the positive control. This was followed by the LF *E. faecium (p<0.01)*, LF *E. faecalis (p<0.01), LF (Blg-Lp) (p<0.01), and LF (Strp-Blg) (p<0.01)* groups.

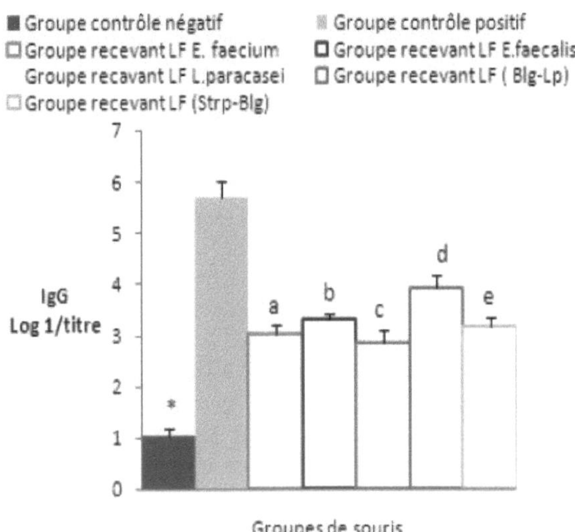

Fig. 32 Serum anti ß-Lg IgG titres in mice at D 35 sensitised intraperitoneally to ß-Lg.
Group **1** negative control; Group **2** positive control; Group **3** intraperitoneally sensitised to ß-Lg followed by oral administration of milk hydrolysates obtained after fermentation by *E. faecium*; Group **4** intraperitoneally sensitised to ß-Lg followed by oral administration of milk hydrolysates fermented by E. faecalis; Group **5** intraperitoneally sensitised to ß-Lg followed by oral administration of milk hydrolysates fermented by *L. paracasei*; Group 6 intraperitoneally sensitised to ß-Lg followed by oral administration of milk hydrolysates fermented by the (Blg-lp) combination. *paracasei* orally; Group **6** sensitised intraperitoneally to ß-Lg and then given hydrolysates of milk fermented with the (Blg-lp) combination orally; Group **7** sensitised intraperitoneally to ß-Lg and then given hydrolysates of milk fermented with the (Strp-Blg) combination orally.

Values are presented as X±SE (n=6)

Values *(p)* represent the comparison of serum IgG titre at D35 between groups of mice sensitised intraperitoneally to ß-Lg and then given milk hydrolysates *.

fermented with the positive control group .(*p)* represents the comparison between group **1** and group **2** *p<0.0001.*[a] p represents the comparison between group 3 and group 2 *p<0.01.*[b] p represents the comparison between group 4 and group **2** *p<0.01.* [c]*p* represents the comparison between group **5** and group **2** *p<0.001.*[d] p represents the comparison between group 6 and group 2 $p<0.01$.^{e} p represents the comparison between group 7 and group 2 $p<0.01$.

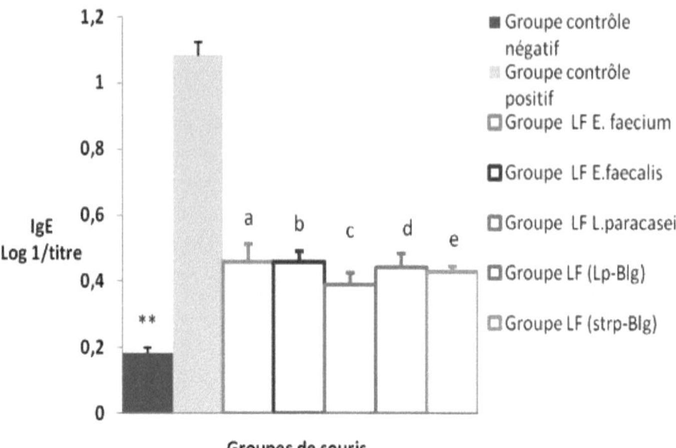

Fig. 33 Anti-ß-Lg serum IgE titres in mice at D35 sensitised intraperitoneally to ß-Lg and then given fermented milk hydrolysates orally.

Group **1** negative control; Group **2** positive control; Group **3** sensitised intraperitoneally to ß-Lg then orally administered a milk hydrolysate fermented by *E. faecium*; Group **4** sensitised intraperitoneally to β -Lg then orally administered a milk hydrolysate fermented by E. faecalis; Group **5** sensitised intraperitoneally to ß-Lg then orally administered a milk hydrolysate fermented by *L. paracasei*; Group **6** sensitised intraperitoneally to ß-Lg then orally administered a milk hydrolysate obtained after fermentation by the (Blg-strp) combination; Group **7** sensitised intraperitoneally to ß-Lg then orally administered a milk hydrolysate fermented by the (Blg-Lp) combination.

Values are presented as X±SE (n=6). **Values (*p*)** represent the comparison of serum IgE titres against ß-Lg at D35 in groups of mice sensitised intraperitoneally to ß-Lg and then given the hydrolysates with the positive control group.** p represents the comparison between group 2 and group 1 $p<0.001$. ^{a}p represents the comparison between group 3 and group 2 $p<0.01$.^{b} p represents the comparison between group 4 and group 2 $p<0.01$.^{c} p represents the comparison between group 5 and group 2 $p<0.01$.^{d} p represents the comparison between group 6 and group 2 $p<0.01$.^{e} p represents the comparison between group 7 and group 2 $p<0.01$.

It should be noted that levels of anti-B-Lg IgE production were undetectable at D0.

2.2.3 Effect of ß-Lg on short circuit current (Isc)

Evaluation of the therapeutic effect was tested in the Ussing chamber in all animals given fermented milk hydrolysates after immunisation with ß-Lg.

Our results show a significant increase in Isc Δ Isc = 24.8µA/cm^2 in the positive control group. In the LF *L. paracasei* group Isc is highly significantly decreased compared to the positive control Δ Isc = 6µAZcm^2 *(p<0.0001)* followed by the LF *faecium* group *p<0.001, LF faecalis p<0.01, LF (*Lp-Blg) p<0.01 and LF (*Strp-Blg) p<0.01* (fig.34)

2.2.4 Measuring conductance (G)

The results obtained in (Table 16) represent the conductance values of tissues from animals that were sensitised intapatitoneally to ß-Lg and then given fermented milk hydrolysates.

Conductance was high in the positive control group, rising from 37.43 ± 1.09 mmho/cm^2 to

62.14 ± 1.31 mmho/cm^2 $p<0.0001$. In the groups of mice given fermented milk hydrolysates, conductance was significantly lower than in the positive control group. The conductance of the negative control group of mice remained unchanged.

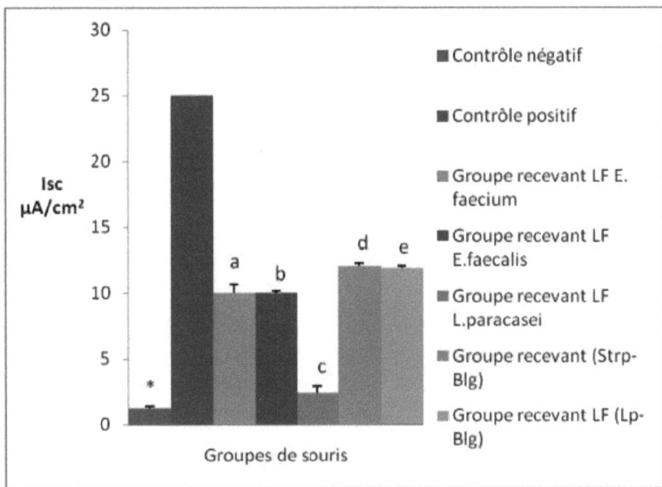

Fig. 34 Effect of ß-Lg on the short-circuit current (Isc) measured in the Ussing chamber on jejunal fragments from mice sensitised intraperitoneally to ß-Lg and then given fermented milk hydrolysates orally.

Group **1** negative control; Group **2** positive control; Group **3** intraperitoneally sensitised to ß-Lg followed by oral administration of *E. faecium* fermented milk hydrolysate; Group **4** intraperitoneally sensitised to ß-Lg followed by oral administration of E. faecalis fermented milk hydrolysate; Group 5 intraperitoneally sensitised to ß-Lg followed by oral administration of L. *faecalis* fermented milk hydrolysate; Group 6 intraperitoneally sensitised to ß-Lg followed by oral administration of E. *faecium* fermented milk hydrolysate. faecalis; Group **5** sensitised intraperitoneally to ß-Lg followed by oral administration of a milk hydrolysate fermented by *L. paracasei*; Group **6** sensitised intraperitoneally to ß-Lg followed by oral administration of a hydrolysate of milk fermented by co-culture (Blg-Lp); Group **7** sensitised intraperitoneally to ß-Lg followed by oral administration of a hydrolysate of milk fermented by co-culture (Strp-BLg).

Values are presented as X±SE (n=6), ***p** represents the comparison of Isc between group 1 and group 2 $p<0.0001$.*a* p represents the comparison of Isc between group 3 and group 2 $p<0.01$.*b* p represents the comparison of Isc between group 4 and group 2 $p<0.01$. *cp* represents the comparison of Isc between group 5 and group 2 $p<0.001$.*d* p represents the comparison of Isc between group 6 and group 2 $p<0.01$.*e* p represents the comparison of Isc between group 7 and group 2 $p<0.01$.

Table 16: Conductance (G) of jejunal fragments from mice sensitised intraperitoneally to ß-Lg and then given oral hydrolysates of fermented milk mounted in Ussing chambers in response to ß-Lg stimulation (60gg/ml).

G

2

(mmho/cm^2

	Before	β Lg	After
	deposit	(60 µgml)	deposit
Negative control group	19,15 ±0,22		20 ± 2,1*
Positive control group	37,43 ± 1,09		62,14 ± 1,31
Group receiving LF *Efaecium*	23,89 ± 1,59		34,05 ± 1,96 [a]
Group receiving LF *E.faecalis*	20± 0,45		32,53± 2,06 [b]
Group receiving *L.paracasei*	28,06± 2,13		37,03 ± 1,5 [c]
Group receiving LF (Blg-Lp)	27,5 ± 1,2		42 ± 2,29[d]
Group receiving LF (Strp-Blg)	25,6 ± 2,72		39,04 ± 2,64[e]

Values represent means ± standard error, X±SE (n=6);

Values *(p)* represent the comparison between conductance (G) at D35 in groups of mice sensitised intraperitoneally to ß-Lg and then given fermented milk hydrolysates, and the positive control group.

**p* represents the comparison between the negative control group and the positive control group *p<0.0001;*[a] p represents the comparison between the group receiving LF *E. faecium* and the positive control group *p<0.001;*[b] p represents the comparison between the group receiving LF *E. faecalis* and the positive control group *p<0.001;*[c] p represents the comparison between the group receiving LF *L. paracasei* and the positive control group *p<0.01;*[d] p represents the comparison between the group receiving LF *(Blg-Lp)* and the positive control group *p<0.01;*[e] p represents the comparison between the group receiving LF (Strp-Blg) and the positive control group *p<0.01.*

Evaluation of the preventive effect of different fermented milk hydrolysates in animals sensitised intraperitoneally to ß-Lg and orally to bovine milk.

We have shown that the hydrolysates obtained after fermentation contain peptide fractions with antioxidant properties.

The modulation of the anaphylactic response to the different hydrolysates was studied *in vitro* in a Ussing chamber on fragments of intestine from mice previously sensitised to the allergens.

Animals sensitised intraperitoneally to ß-Lg

We observed a low production of serum anti-ß-Lg IgG and IgE in all groups of animals treated with hydrolysates compared with the positive control. However, this reduction was more marked in the *LF paracasei* group.

The hydrolysates obtained under these conditions can have the following properties: on the one hand, they can reduce the allegenic epitopes of milk proteins and, on the other hand, the peptides resulting from hydrolysis can also have antioxidant properties.

This low production of serum IgG and IgE can be explained by the fact that during the fermentation process, derivatives are released after hydrolysis of milk proteins by the peptidases and proteases of lactic bacteria. These derivatives can be considered as peptides with modulating and regulating potential in the body (Ganjam et *al.*, 1997;Meisel and Bockelmann, 1999; Takano, 2002; Vinderola et *al.*, 2007). Studies have shown that immunomodulatory compounds are released after fermentation of milk by the strain *Lactobacillus helveticus* R389 (Matar et *al.*, 2001). The peptides thus produced in the same strain induce a protective humoral immune response following infection with E. coli (LeBlanc et *al.*, 2004).

The reduction in the number of epitopes and consequently the reduction in the antigenicity/allergenicity of hydrolysed proteins has already been observed by several authors (Wroblewska et *al.*, 1995; Cross et *al.*, 2001; Bertrand-harb et *al.*, 2003; Nentwich et *al.*, 2004). Other authors (Kleber et *al.*, 2006; Guanhao et *al.*, 2013) have also shown that strains of lactic acid bacteria reduce the antigenic responses of milk proteins. Our results suggest that the proteolytic activity observed in the strains studied may lead to a reduction in the antigenicity of milk proteins.

In our work, we observed the persistence of anti-ß-Lg IgE antibodies in animals receiving hydrolysates, but to a lesser extent than in the positive control. This suggests that epitopes recognised by IgE persist in the hydrolysates. A study on different *L. helveticus* strains conducted by Ehn et *al* (2005) on milk whey showed a hydrolysis of more than 80% for ß-Lg, but the epitopes recognised by IgE were not affected. Our results suggest, as reported by Kleber et *al* (2006), that the proteolytic activity in the fermentation process is partial and does not allow total disappearance of the epitopes recognised by IgE.

The effect of ß-Lg on short circuit current (Isc) was assessed on jejunal fragments from mice after the *in vitro* Ussing chamber challenge test.

Our results show that stimulation with the allergen causes a significant increase in Isc in the positive control group. Numerous studies have also shown a significant increase in short-circuit current following contact of the sensitising antigen with Ussing chamber-mounted intestinal fragments (Saidi, 1995; Berin et *al.*,1997; Berin et *al.*,1998; Yang et *al.*,2000, Addou et *al.*,2004; Zellal et *al.*,2011). This increase in Isc corresponds to an electrogenic secretion of Cl$^-$ triggered by the ß-Lg sensitising antigen (Devor et *al.*, 2000; Yang et *al.*, 2000; Zellal et *al.*, 2011).

In general, the groups that received the different hydrolysates showed a significantly lower response to allergen stimulation than the positive control. This effect was more marked in the LF *L. paracasei* group.

In parallel with Isc, there is also an increase in conductance, which is thought to be one of the consequences of allergens passing through the intestinal epithelium (Kheroua et *al.*, 1987; Heyman et *al.*, 1990;Saidi et *al.*, 1995;Brandt et *al.*, 2003). The conductance of the tissues studied after stimulation by the sensitising antigen increases significantly in the positive control group. This corresponds to an increase in intestinal permeability. Tissue conductance in the groups of mice given fermented milk hydrolysates was significantly lower than that observed in the positive control group.

In the light of these results, the hydrolysates studied could modulate the anaphylactic response at different levels. The *L. paracasei* strain appears to be an interesting candidate as it very significantly reduces serum immunoglobulins and at the same time reduces short circuit current and conductance. Kume et *al*, (2014) demonstrated that oral ingestion for 14 days of a diet containing whey peptides and fermented dairy products had a protective effect in rats suffering from disorders of the grele intestine.

In order to determine the nature of the increase in Isc, tissues from all the groups studied were treated with furosemide, a diuretic known to act rapidly on the inhibition of Cl/Na/K cotransporters at the basolateral membrane of the enterocyte, leading to inhibition of chlorine secretion by CFTR channels (Devor et *al.*, 2000).

Our results showed that furosemide inhibited the short circuit current in all groups of animals (control and experimental). The increase in Isc observed in our work corresponds to chlorine secretion.

These results are in agreement with Negaoui et *al* (2009), who showed that no significant change was observed after the addition of furosemide.

Mice orally sensitised to cow's milk

The results show that the positive control group produced highly significant levels of anti-ß-Lg serum IgG and IgE. For the groups of mice that received the fermented milk hydrolysates, we observed a lower production of serum IgG and IgE compared with the positive control.

Numerous studies mention an anti-allergic effect of fermented dairy products, but few authors really demonstrate whether this effect is linked to the hydrolysis of allergic epitopes or to a modulation of the immune response by lactic acid bacteria (Cross et *al.*, 2001). Numerous studies have shown that fermentation of bovine milk by lactic acid bacteria induces a significant reduction in the antigenicity of a-la, ß-Lg and bovine serum albumin (Jedrychowski and Wroblewska, 1999; Ckekroun et *al.*, 2006; Belkaaloul et *al.*, 2012). Studies have long reported that food antigen fractions combine with antibodies in the intestinal lumen, forming immune complexes that can prevent the absorption of antigenic molecules and react with helper T cells and suppressor T cells and subsequently regulate the production of IgG, IgA and IgE (Murphy and Walker, 1991; Sorenson et *al.*, 1993). Vinderola et *al* (2007) studied the effects of the ingestion of milk fermented with *Lactobacillus helveticus*

R389 at pH 6 for 2, 5 and 7 days in the intestinal mucosa of healthy BALB/c mice. For all durations of administration, an increase in the number of cells producing IgA, IL-10, IL-2 and IL-6 was demonstrated. Total IgA production in the lumen of the large intestine was stimulated in mice administered the product for 2 days, as was IL-6 production by epithelial cells after 7 days of ingestion.

Evaluation of the therapeutic effect of fermented milk hydrolysates in mice sensitised intraperitoneally to ß-Lg.

The study of the evaluation of the therapeutic effect in the Ussing chamber showed that in all the groups receiving the fermented milk hydrolysates, there was a decrease in Isc compared with the positive control. For the LF *L. paracasei* group, the decrease in Isc was more marked. The effect observed is probably linked to the antioxidant properties of the peptides produced during fermentation. Indeed, it has been shown that bioactive peptides with antioxidant activity modulate the immune response (Korhonen & Pihlanto, 2006; Phelan et *al.*, 2009b). They also reported that bacterial hydrolysis produces bioactive peptides. Several studies in the literature have addressed the immunomodulatory activity induced by milk fermented with *L. Helveticus*. A series of publications by Matar's group demonstrated the immunomodulatory effect of milk fermented with *L. Helveticus R389* and identified the peptides involved (Matar, et *al.*, 1996; Matar, et *al.*, 2001; LeBlanc et *al.*, 2002; LeBlanc et *al.*, 2004). A subsequent study was therefore carried out in healthy mice by Hébert-Leclerc, (2011) to assess the immunomodulatory effects of a peptide derived from the hydrolysis of β-Lg in purified form, following oral ingestion for 7 days. This peptide caused a significant increase in IgA levels measured in the faeces, but did not alter serum IgA levels. In addition, IL-4 production by splenocytes was stimulated. Orally administered ß-LG 27 f 96-99 peptide therefore appears to have a stimulatory effect on the adaptive immune response of healthy mice.

A study by Santos et *al* (2014) in allergic children showed that after oral challenge with fermented milk hydrolysates, oral tolerance to cow's milk proteins was established. Our results suggest that oral tolerance was established in groups that received milk hydrolysates Santos et *al*, (2014), Bennett et *al*, (2005), Jing & Kitts, (2004), Migliore- Samour et *al*, (1989); Phelan et *al*, (2009) and Sandré et *al*. (2001) have reported that milk contains numerous peptides that affect the immune system via cellular functions and play an important role in the immunomodulation of casein hydrolysates.

2.3 Histological study

2.3.1 Mucous membranes of mice orally treated with fermented milk hydrolysates and then orally sensitised to cow's milk

Our results show that in the positive control group, the villi are atrophied with pseudostratified epithelium and the chorion is markedly inflamed (fig. 35 D).

In the groups receiving hydrolysates (fig. 35 B, C, E, F, G) we observed minor histological damage with weaker inflammatory signs and villi that were more or less enlarged. Fermented milk hydrolysates appear to have a protective effect on the intestinal epithelium and reduce the histological lesions caused by cow's milk proteins. The villi of the negative control group were long with unistratified epithelium (fig.35 A). Similar results were observed in rabbits by Addou et *al* (2004). Probiotics have also been reported to play a role in protecting the architecture of the intestinal mucosa (Frick et *al.,* 2007).

2.3.2 Mucous membranes of mice given fermented milk hydrolysates and then intraperitoneally sensitised to ß-Lg

(fig. 36 I) shows a histological section of the mucosa of the positive control group. It is characterised by villi lined with pseudostratified epithelium, containing cubic cells with dystrophic nuclei, and in the chorion the inflammatory infiltrate was extremely marked. Histological sections of intestinal mucosa from groups of mice receiving hydrolysates are shown in (figs. 36 J, K, L, M, N). The villi are less atrophied. Our results suggest that the hydrolysates studied in our protocol protect the architecture of the intestinal mucosa.

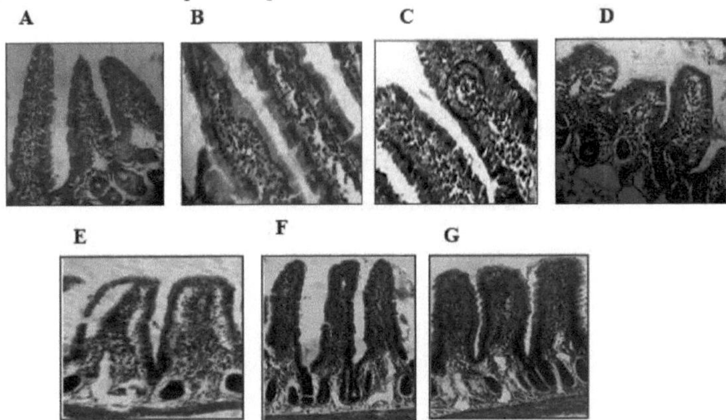

Fig.35 Microscopic observation of jejunal fragments from mice at D35 receiving fermented milk hydrolysates and then orally sensitised to cow's milk (X40).

A. Jejunal fragment from a mouse in the negative control group;

B. Jejunal fragment of a mouse from the group receiving LF *L. paracasei* orally and then sensitised to cow's milk orally;

C. Jejunal fragment from mice in the group receiving LF *E. faecium* orally and then

69

sensitised to cow's milk orally;

D. Jejunal fragment from a mouse in the positive control group;

E. Jejunal fragment from mice in the group receiving LF *E. faecalis* orally and then sensitised to cow's milk orally;

F. Jejunal fragment from mice in the group receiving LF *(Blg-Lp)* orally and then sensitised to cow's milk orally;

G. Jejunal fragment from mice in the group receiving LF (*Strp-Lp*) orally and then sensitised to cow's milk orally

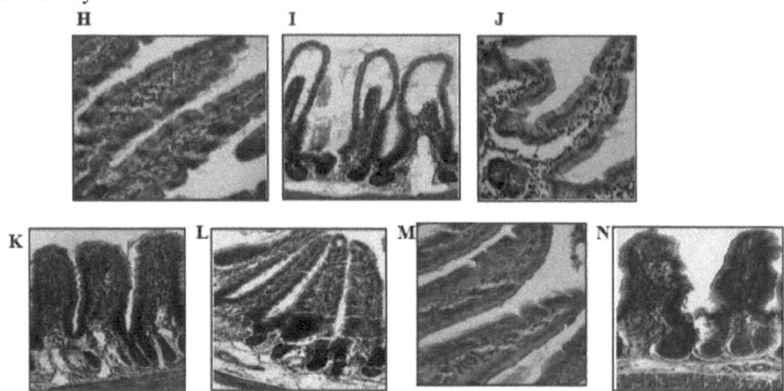

H.

I. **Fig. 36 Microscopic observation of jejunal fragments from mice at D35 receiving hydrolysates of fermented milk orally and then sensitised intraperitoneally to ß-Lg (X40).**

J. Jejunal fragment from a mouse in the negative control group;

K. Jejunal fragment from a positive control mouse sensitised intraperitoneally to ß-Lg;

L. Jejunal fragment from mice in the group receiving LF *E. faecium* orally and then sensitised to ß-Lg intraperitoneally;

M. Jejunal fragment from mice in the group receiving LF *(Blg-Lp)* orally and then sensitised to ß-Lg intraperitoneally;

N. Jejunal fragment from mice in the group receiving LF *L. paracasei* orally and then sensitised to ß-Lg intraperitoneally;

O. Jejunal fragment from mice in the group receiving LF *(Blg-Lp)* orally and then sensitised to ß-Lg intraperitoneally.

P. Jejunal fragment from mice in the group receiving LF *E. faecalis* orally and then sensitised to ß-Lg intraperitoneally.

Conclusion

Food proteins are a source of peptides with various biological activities, such as antihypertensive activity, immunostimulant activity and antioxidant activity. These peptides are produced *in vivo/in vitro* and during food processing. Bioactive peptides are widely present in cheeses and fermented milks. Proteolytic enzymes from lactic acid bacteria also contribute to the production of bioactive peptides.

In this work, we addressed two important points in order to determine whether the use of degradation products from bacterial hydrolysis of cow's milk proteins by selected lactic acid bacteria is a good alternative for protecting the intestinal barrier. The first part of the work involved assessing the performance of the strains used during fermentation. All the strains had a good fermentation profile, since acidification of the medium was achieved rapidly. Electrophoretic analysis by SDS-PAGE of the hydrolysates of the milk fermented by these strains showed a degradation of milk proteins, particularly caseins, with the appearance of peptides of different molecular weights. The proteases responsible for this hydrolysis are mainly metalloproteases. HPLC chromatographic analysis showed that hydrophobic peptides were generated at different linear gradients and retention times for each hydrolysate. Evaluation of the antioxidant capacity of the hydrophobic peptide fractions of the hydrolysates showed that the peptides produced by the *L.paracasei* strain have a more marked antioxidant activity. This suggests that *L.paracasei* may be a good candidate for producing fermented milks with functional antioxidant properties. Secondly, an *ex-vivo* Ussing chamber evaluation of the preventive effect showed that prior administration of hydrolysates to mice then sensitised intraperitoneally to B-Lg and orally to cow's milk reduced the level of production of anti ß-Lg IgG and IgE immunoglobulins. The therapeutic effect study also showed that administration of hydrolysates attenuated the immune response. The preventive and therapeutic properties were more marked with the *L.paracasei* strain. This suggests that during the fermentation process, peptide derivatives are released following hydrolysis of milk proteins by the peptidases and proteases of lactic acid bacteria. These derivatives may be peptides with modulating and regulatory potential in the body. *L.paracasei* appears to be of interest
since it very significantly reduces serum immunoglobulins and at the same time reduces short-circuit current and conductance in animals sensitised to the major sensitising antigen in bovine whey, β-Iactoglobulin.

We can suggest that the hydrolysates obtained may have the following properties: on the one hand, they may reduce the allegenic epitopes of milk proteins, reduce inflammation and protect the integrity of the intestinal barrier, and on the other hand, the peptides resulting from hydrolysis may also have antioxidant properties.

In the future, it would be interesting to assess the other effects on human health of the bioactive peptides produced by hydrolysis by the strains selected in this work, and a complete analysis of the peptide profile would be important for sequencing the peptides involved in this physiological effect.

References

1. Addou S., D. Tomé, O. Kheroua, D. Saidi, Parenteral immunization to β-Iactoglobulin modifies the intestinal structure and mucosal electrical parameters in rabbit, *International. Immunopharmacology*. 2004; 4:1559-1563.
2. Adel-Patient K., Bernard H., Wal J.-M. Fate of allergens in the digestive tract. *Revue Française d'Allergologie et d'Immunologie Clinique*, 2008 ; 48: 335-343.
3. Adson A., Raub T.J., Burton P.S., Barsuhn C.L., Hilgers A.R., Audus K.L., and Ho N.F. Quantitative approaches to delineate paracellular diffusion in cultured epithelial cell monolayers. *Journal of Pharmaceutical Sciences*, 1994; 83:1529-1536.
4. Ahmadova A., Dimov S., Ivanova I., Choiset Y., Chobert J.-M., Kuliev A., Haertlé T. Proteolytic activities and safety of use Enterococci strains isolated from traditional Azerbaijani dairy products. *European Food Research and Technology*, 2011; 233: 131-140.
5. Aloglu H. S. and Oner Z. Determination of antioxidant activity of bioactive peptide fractions obtained from yogurt. *Journal of Dairy Science*, 2011 ; 94: 5305-5314.
6. Amiot J., Fournier S., Lebeuf Y., Paquin P., Simpson R. Chapitre 1 : Composition, propriétés physicochimiques, valeurs nutritive, qualité technologique et techniques d'analyse du lait. In : Science et Technologie du lait ; transformation du lait.C. L.Vignola, ed. Presses Internationales Polytechniques, Montréal, Québec, p.1-73.2002.
7. Ammor M., Flórez A., Mayo B. Antibiotic resistance in non-enterococcal lactic acid bacteria and bifidobacteria. *Food Microbiology*, 2007; 24: 559-570.
8. Anema S. G. Effect of milk solids concentration on whey protein denaturation, particle size changes and solubilization of casein in high-pressure-treated skim milk. *International Dairy Journal*, 2008; 18: 228-235.
9. Apostolidis E., Kwon Y.I.I., Shetty K. Potential of cranberry-based herbal synergies for diabetes and hypertension management-ment, *Asia Pacific Journal of Clinical Nutrition*, 2006; 15:433-441.
10. Argyri A. A., Zoumpopoulou G., Karatzas K.A. G., Tsakalidou E., Nychas G.J. E., Panagou E. Z., Tassou C. C. Selection of potential probiotic lactic acid bacteria from fermented olives by in vitro tests. *Food Microbiology*, 2013; 33:282-291.
11. Arizcum C., Barcina Y. & Torre P. Identification and characterisation of proteolytic activity of *enterococcus* spp. Isolated from raw milk and Roncal and idiazabal cheese. *Lait*, 1997; 77: 729-736.
12. Aslim B., Yuksekdag Z.N., Sarikaya E., Beyatli Y. Determination of the bacteriocin-like substance produced by some lactic acid bacteria isolated from Turkish dairy products, LWT-FoodScience *and Technology*,2005; 38: 691-694.
13. Axelsson L. Lactic acid bacteria: classification and Physiology.in Lactic acid bacteria. Microbiological and functional aspects. New York: Marcel Dekker, Inc, 2004, p.1-66.
14. Barbosa J., Ferreira V., Texeira P. Antibiotic susceptibility of Enterococci isolated from traditional fermented meat products. *Food Microbiology*, 2009; 26: 527-532.
15. Barnig C., Schulmeister U., Swoboda I., Bessot J .C. Spitzauer S., Pauli G. Cow's milk protein allergy without associated sheep's milk allergy in adults. *Revue Française d'Allergologie et d'Immunologie Clinique*, 2005 ; 45:608-11.
16. Batdorj B., Dalgalarrondo M., Choiset Y., Pedroche J., Metro F., Prevost H., Chobert JM., Haertlé T. Purification and characterization of two bacteriocins produced by lactic acid bacteria isolated from Mongolian airag. *Journal of Applied Microbiology*, 2006; 101: 837-

848.

17. Belkaaloul K, Chekroun A, Chobert J.-M., Haertlé T., Saidi D. and O. Kheroua. Prevention of Gut Mucosa Inflammation by Two Co-cultures of Lactobacillus plantarum-Bifidobacterium longum and Streptococcus thermophilus-Bifidobacterium longum. *Journal of Food Science and Engineering.*2012; 2: 685-690.

18. Belut D., Moneret-Vautrin D.A., Nicolas J.P., Grilliat J.P. IgE levels in intestinal juice. *Digestive Diseases and Sciences*, 1980; 25:323-32.

19. Ben Belgacem Z., Abriouel H., Ben Omar N., Lucas R., Martínez-Canamero M., Gálvez A., Manai M. Antimicrobial activity, safety aspects, and some technological properties of bacteriocinogenic *Enterococcus faecium* from artisanal Tunisian fermented meat. Food Control, 2010; 21(4): 461-470.

20. Ben omar N., Castro A., Lucas R., Abriouel H., Yousif N. M., Franz C., Holzalpef M.,Pérez-Pulido W. H. R., Martinez-Cañamero M., Gálvez A. Functional and safety aspects of enterococci isolated from different Spanish foods. *Systematic and Applied Microbiology*, 2004; 27:118-130.

21. Bennett L. E., Crittenden R., Khoo E., & Forsyth S. Evaluation of immunemodulatory properties of selected dairy peptide fractions. *Australian Journal of Dairy Technology*, 2005; 60:106-109.

22. Berin M.C., Kiliaan A.J., Yang P.C., Groot J.A., Taminiau J.A., Perdue M.H. Rapid transepithelial antigen transport in rat jejunum: impact of sensitization and the hypersensitivity reaction. Gastroenterology, 1997; 113:856-864.

23. Berin M.C., Kiliaan A.J., Yang P.C., Groot J.A.,Kitamura Y., Perdue M.H. The influence of mast cells on pathways of transepithelial antigen transport in rat intestine. *The journal of immunology*, 1998; 161: 2561-2566.

24. Bertrand-Harb C., Ivanova I., Dalgalarrondo M., Haertlé T. Evolution of β-lactoglobulin and α-lactalbumin content during yoghurt fermentation. *International Dairy Journal*, 2003; 13: 39-45.

25. Besler M., Steinhart H., Paschke A. Stability of food allergens and allergenicity of processed foods. *Journal of Chromatography,*2001; 756 : 207- 228.

26. Bidat E. Food allergy in children. *Archives de pédiatrie*, 2006; 13 :1349-1353.

27. Bischoff S.C. & Crowe S.E. Gastrointestinal food allergy: new insights into pathophysiology and clinical perspectives. *Gastroenterology,* 2005; 128, 1089-1113.

28. Brandt E.B., Strait R.T., Hershko D., Wang Q., Muntel E.E., Scribner T.A., Zimmermann N., Finkelman F.D., Rothenberg M.E. Mast cells are required for experimental oral allergen-induced diarrhea. *Journal of Clinical Investigation*, 2003; 112: 1666-1677.

29. Brandtzaeg P. The gut as communicator between environment and host: Immunological consequences. *European Journal of Pharmacology*, 2011; 668(1): 16-32.

30. Broadbent J., Steele J. Cheese flavor and the genomics of lactic acid bacteria. American Society for Microbiology. *News*, 2005; 71: 121-128.

31. Bu G., Luo Y., Chen F.,K. Liu, Zhu T. Milk processing as a tool to reduce cow's milk allergenicity: a mini-review. *Journal of Dairy Science and technology*, 2013 ; 93:211223.

32. Callewaert R. and de Vuyst L. Bacteriocin production with *Lactobacillus amylovorus* DCE 471 is improved and stabilized by fed-batch fermentation. *Applied and Environmental Microbiology*, 2000; 66: 606-613.

33. Carr F. J., Chill D., Maida N. The lactic acid bacteria: A literature survey, *Critical Reviews in Microbiology*, 2002; 28 (4): 281-370.

34. Chatchatee P., Jarvinen K., Bardina L., Vila L., Beyer K., Sampson H.A. Identification of IgE and IgG binding epitopes on beta and kappa-casein in cow's milk allergic patients. *Clinical & Experimental Allergy*, 2001; 31: 1256-1262.

35. Chirdo F.G., Millington O.R., Beacock-Sharp H., Mowat A.M. Immunomodulatory dendritic cells in intestinal lamina propria. *European Journal of Immunology*, 2005; 35: 1831-40.

36. Christensen J.E., Dudley E.G., Pederson J.A., Steele J.L. Peptidases and amino acid catabolism in lactic acid bacteria. *Antonie Van Leeuwenhoek*, 1999; 76: 217-246.

37. Cintas L.M., Herraz C., Hernandez P.E., Casaus M.P., Nes I.F., et Hernandez P.E. Bacteriocin of lactic acid bacteria. *Food Science and Technology International*, 2000; 7(4): 281-305.

38. Ckekroun A., Bensoltane A., Kheroua O., Saidi D. Biotechnological characteristics of fermented milk by bacterial associations of the st strains Streptococcus, Lactobacillus and bifidobacterium. *Egyptian Journal of Basic and Applied Sciences,* 2006; 21(2b): 583-598.

39. Clare D. A., & Swaisgood H. E. Bioactive milk peptides: a prospectus. Journal of Dairy Science, 2000; 83: 1187-1195.

40. Cocco R. R., J€arvinen K. M., Sampson H. A., & Beyer K. Mutational analysis of major, sequential IgE-binding epitopes in αS1-casein, a major cow's milk allergen. *Journal of Allergy and Clinical Immunology*, 2003; 112: 433-437.

41. Cole Z. A., Clough G. F. & Church M. K. Inhibition by glucocorticoids of the mast cell-dependent weal and flare response in human skin in vivo. *British Journal of Pharmacology*, 2001; 132: 286-292.

42. Çon A. H. and Gökalp A.Y. Production of bacteriocin-like metabolites by lactic acid cultures isolated from sucuk samples. Meat Science, 2000; 55(1): 89-96.

43. Considine T., Patel H. A., Anema S. G., Singh H., and Creamer L. K. Interactions of milk proteins during heat and high hydrostatic pressure treatments-A review. *Innovative Food Science and Emerging Technologies*, 2007; 8: 1-23.

44. Conway P.L., Gorbach S.L., Goldin B.R. Survival of lactic acid bacteria in the human stomach and adhesion to intestinal cells. *Journal of Dairy Science,* 1987; 70(1): 1-12.

45. Cross M., Stevenson L., Gill H. Anti-allergy properties of fermeted foods: an important immunoregulatorymechanism of lactic acid bacteria? *International Immunopharmacollogy*, 2001; 1:891-901.

46. De Boissieu D, Dupont C. IgE-mediated cow's milk allergy. *Archives de pediatrie*, 2006; 13:1283-1284.

47. De Leo, F., Panarese, S., Gallerani, R., & Ceci, L. R. Angiotensin converting enzyme (ACE) inhibitory peptides: production and implementation of functional food. *Current Pharmaceutical Design*, 2009; 15: 3622- 3643.

48. De Man J. C., Rogosa M., Sharpe E.M. A medium for the cultivation of lactobacillii. *Journal of Applied Bacteriology*, 1960; 23: 130-135.

49. De Vuyst L. and Vandamme E. J. (1994). Antimicrobial potential of lactic acid bacteria. In L. de Vuyst, & E. J. Vandamme (Eds.), Bacteriocins of lactic acid bacteria: microbiology, genetics and applications (pp. 91e142). London, UK: Blackie Academic & Professional.

50. Desjeux J.-F., Heyman M., GRASSET E. Membrane transport systems, genetics and nutrition; the example of congenital abnormalities of intestinal transport in children. *Reproduction Nutrition Development*,1984; 24(5B) :785-792.

51. Desmazeaud M. L'état des connaissances en matière de nutrition sur les bactéries

lactiques. *Le lait*, 1983; 63: 286-310.

52. Devor D.C., Bridges J.R., Pilewski J.M. Pharmacological modulation of ion transport across wild type and Delta F508 CFTR-expressing human bronchial epithelia. *American Journal of Physiology and cell physiology;* 2000; 279: 461- 479.

53. Diaz Muniz I., Banavara D., Budinich M., Rankin S.A., Dudley E.G. Steel J.L. Lactobacillus casei metabolic potential to utilize citrate as an energy source in ripening cheese: a bioinformatics approach. *Journal of Applied Microbiology*, 2006; 101: 872882.

54. Donkor O. N. (2007). Influence of probiotic organisms on release of bioactive compounds in yoghurt and soy yoghurt. Ph.D Thesis, Victoria University, Melbourne, Australia.

55. Du Toit M., Franz C.M., Dicks L.M., Schillinger U., Haberer P., Warlies B., Ahrens F., Holzapfel W.H. Characterisation and selection of probiotic lactobacilli for a preliminary minipig feeding trial and their effect on serum cholestrerol levels, faeces pH and faeces moisture content. *International Journal of Food Microbiology*, 1998; 40: 93- 104.

56. Dunne C., O'Mahony L., Murphy L., Thornton G., Morrissey D., O'Halloran S., Feeney M., Flynn S., Fitzgerald G., Daly C., Kiely B., O'Sullivan G.C., Shanahan F., Collins J.K., In vitro selection criteria for probiotic bacteria of human origin: correlation with in vivo findings. *American Journal of Clinical Nutrition*, 2001; 73 (2):386-392.

57. Ehn B., Allmere T., Telemo E., Bengtsson U. Modification of IgE Binding to α-lactoglobulin of genetic variants and influence of purification method on secondary structure. *Milwissenschaft*, 1990; 45: 694-698.

58. Ejtahed H. S., Mohtadi-Nia J., Homayouni-Rad A., Niafar M., Asghari-Jafarabadi M., and Mofid V. Probiotic yogurt improves antioxidant status in type 2 diabetic patients.. *Nutrition,* 2012; 28: 539-543.

59. Elfahri K.R., Donkor O.N., Vasiljevic T. Potential of novel *Lactobacillus helveticus* strains and their cell wall bound proteases to release physiologically active peptides from milk proteins. *International Dairy Journal*, 2014; 38(1): 37-46

60. El mecherfi K.E.D. Evaluation of the immunoreactivity and allergenicity of bovine lactoproteins after enzymatic hydrolysis combined with microwave treatment. Interest of allergen microarrays in the multi-detection of specific IgE responses in children poly-sensitized to cow's milk proteins. 2012. PhD thesis.

61. El Mecherfi K.E., Rouaud O., Curet S., Negaoui H., Chobert J.-M., Kheroua O., Saidi D. and Haertlé T. Peptic hydrolysis of bovine beta-lactoglobulin under microwave treatment reduces its allergenicity in an *ex vivo* murine allergy model. *International Journal of Food Science & Technology*, 2015; 50(2): 356-364.

62. El Shafei H.A., Adel-sabour H., Ibrahim N., and Mostefa Y.A. Isolation screening and characterisation of the bacteriocin producing lactic and bacteria isolated from traditionally fermented. *Food Microbiology Research,* 2000; 154(4):321-331.

63. El Soda M., Ahmed N., Omran N., Osman G. and Morsi A. Isolation, identification and selection of lactic bacteria cultures forcheese-making. *Emirates Journal of Food and Agriculture*, 2003; 15:51-71.

64. El-Baradei G., Delacroix-Buchet A. and Oiger J.C. Bacterial biodiversity of traditional Zabady fermented milk. *International Journal of Food Microbiology*, 2008; 121:295301.

65. El-Ghaish S., Dalgalarrondo M., choiset Y., Sitohy M., Ivanova I., Haertlé T. Chobert J.-M. Screening of strains of lactococci isolated from Egyptian dairy products for their proteolytic activity. *Food Chemistry*, 2010; 120:758-764.

66. El-Ghaish S., Hadji-Sfaxi I., Ahmadova A., Choiset Y., Rabesona H., Sitohy M., Haertlé T., & Chobert J.-M. Characterization of two safe *Enterococcus* strains producing enterocins isolated from Egyptian dairy products. *Beneficial Microbes*, 2011; 2:15-27.

67. Elias R.J., Kellerby S.S., Decker E.A. Antioxidant activity of proteins and peptides. *Critical Reviews in Food Science and Nutrition*, 2008; 48 (5): 430-441.

68. Erdmann K., Cheung B. W. Y., Schröder H. The possible roles of food derived bioactive peptides in reducing the risk of cardiovascular disease. *The Journal of Nutritional Biochemistry*, 2008; *19*: 643-654.

69. Exterkate F. A., Alting A. C., & Bruinenberg P. G. Diversity of cell envelope proteinase specificity among strains of Lactococcus lactis and its relationship to charge characteristics of the substrate-binding region. *Applied and Environmental Microbiology*, 1993; 59, 3640-3647.

70. Fernandez-Espla M.D., Garault P., Monnet V., Rul F. Streptococcus thermophilus cell wall-anchored proteinase: release, purification, and biochemical and genetic characterization. *Applied and Environmental Microbiology*, 2000; 66:4772-4778.

71. Fiocchi A., Brozek J., Schunemann H., Bahna S., Von Berg A., Beyer K., Bozzola M., Bradsher J., Compalati E., Ebisawa M., Guzman M. A., Li H., Heine R. G., Keith P., Lack G., Landi M., Martelli A., Rancé F., Sampson H., Stein A., Terracciano L., and Vieths S. World Allergy Organization (WAO) Diagnosis and Rationale for Action against Cow's Milk Allergy (DRACMA) Guidelines. *Pediatric Allergy and Immunology*, 2010; 21: 1-125.

72. Fira D., Kojic M., Banina A., Spasojevic I., Strahinic I. and Topisirovic L. Characterization of cell envelope associated proteinases of thermophilic lactobacilli. *Journal of Applied Microbiology*, 2001; 90: 123-130.

73. Fox A.T., Thomson M., Adverse reactions to cow's milk. *Pediatrics and child health*, 2007; 17:288-294.

74. Fox P., Singh T., McSweeney P. Biogenesis of flavor compounds in cheese, in: Malin E.L., tunick M.H. (Eds), Chemistry of Structure/Function Relationships in cheese, Plenum Press. New York, 1993; 59-98.

75. Fox P., Wallace J. Formation of flavour compounds. *Advances in Applied Microbiology, 1997;* 45: 17-85.

76. Franz C. M. A. P., Stiles M. E., Schleifer K. H., & Holzapfel W. H. Enterococci in foods e a conundrum for food safety. *International Journal of Food Microbiology*, 2003; 88:105-122.

77. Freitas A. C., Pintado A. E., Pintado M. E., & Malcata F. X. Organic acids produced by lactobacilli, enterococci and yeasts isolated from picante cheese. *European Food Research and Technology*, 1999; 209: 434-438.

78. Frick J.S., Fink K., Kahl F., Niemiec M.J., Quitadamo M., Schenk K., et al, Identification of commensal bacterial strains that modulate *Yersinia enterocolitica* and dextran sodium sulfate-induced inflammatory responses: Implications for the development of probiotics, Infect. Immun, 2007; 75: 3490-3497.

79. Fritsche R. Role for technology in dairy allergy. *Australian Journal of Dairy Technology,* 2003; 58:89-91.

80. Galli S.J. Mast cells and basophils.*Current Opinion in Hematology*, 2000; 7:32-39.

81. Galli S.J. Kalesnikoff J., Grimbaldeston M.A., Piliponsky A.M., Williams C.M., Tsai M. Mast cells as "tunable" effector and immunoregulatory cells: recent advances. *Annuals Review of Immunology*, 2005; 23: 749-786.

82. Ganjam L.S., Thornton W.H., Marshall R.T., Macdonald R.S. Antiproliferative effects of

yogurt fractions obtained by membrane dialysis on cultured mammalian intestinal cells. *Journal of Dairy Science*, 1997; 80: 2325-2329.

83. Gardini F., Martuscelli M., Caruso M. C., Galgano F., Crudele M. A., Favati F., Guerzoni M. E., Suzzi G. Effects of pH, temperature and NaCl concentration on the growth kinetics, proteolytic activity and biogenic amine production of *Enterococcus faecalis*. *International Journal of Food Microbiology*, 2001; 64: 105-117.

84. Gilbert C., Atlan D., Blanc B., Portailer R., Germond J. E., Lapierre L., Mollet B. A new cell surface proteinase: Sequencing and analysis of the prtB gene from Lactobacillus delbrueckii subsp. bulgaricus. Journal of Bacteriology, 1996; 178, 3059-3065.

85. Gilbert C., Blanc B., Frot-Coutez J., Portalier R., Atlan D. Comparaison of cell surface proteinase activities within the Lactobacillus genus. *Journal of Dairy Science*, 1997; 64: 561-571.

86. Gill H. S., Doull F., Rutherfurd K. J., Cross M. L. Immunoregulatory peptides in bovine milk. *British Journal of Nutrition*, 2000; 84(1): 111-117.

87. Giraffa G. Functionality of enterococci in dairy products, *International Journal of Food Microbiology,* 2003; 88: 215-222.

88. Gotcheva V., Pandiella S., Angelov A., Roshkova Z. Microflora identification of the Bulgarian cereal-based fermented beverage boza. *Process biochemistry*, 2000; 36(1/2): 127-130.

89. Gousia P., V. Economou H. Sakkas S. Leveidiotou and C. Papadopoulou. Antimicrobial resistance of major foodborne pathogens from major meat products. *Foodborne Pathogens and Disease,* 2011 ; 8: 27-38.

90. Guanhao B. , Yongkang L. , Fusheng C., Kunlun L., Tingwei Z. Milk processing as a tool to reduce cow's milk allergenicity: a mini-review. *Dairy Science & Technology*, 2013; 93:211-223.

91. Gomez-Ruiz J. A., Ramos M., and Recio I. Angiotensinconverting enzyme-inhibitory peptides in Manchego cheeses manufactured with different starter cultures. *International Dairy Journal*, 2002; 12: (8), 697-706.

92. Hafeez Z., Cakir-Kiefer C. Girardet J.-M., Jardin J., Perrin C., Dary A., Miclo L. Hydrolysis of milk-derived bioactive peptides by cell-associated extracellular peptidases of Streptococcus thermophilus. *Applied Microbiology and Biotechnology*, 2013; 97: 9787-9799.

93. Haquea E., Chanda R., & Kapilab, S. Biofunctional properties of bioactive peptides of milk origin. *Food Reviews International*, 2009; 25: 28-43.

94. Hartmann R., & Meisel H. Food-derived peptides with biological activity: from research to food applications. *Current Opinion in Biotechnology*, 2007; 18:163- 169.

95. Hata I., Ueda J., and Otani H. Immunostimulatory action of a commercially available casein phosphopeptide preparation, CPPIII, in cell cultures. *Milchwissenschaft,* 1999; 54(1): 3-7.

96. Hayes M., Ross R. P., Fitzgerald G. F., Hill C., and Stanton C. Casein-derived antimicrobial peptides generated by Lactobacillus acidophilus DPC6026. *Applied and Environmental Microbiology*, 2006; 72: 2260- 2264.

97. Hayes M., Stanton C., Slattery H., O'Sullivan O., Hill C., Fitzgerald G. F., Ross R.P. Casein fermentate of Lactobacillus animalis DPC6134 contains a range of novel propeptide angiotensin-converting enzyme inhibitors. *Applied and Environmental Microbiology*, 2007; 73: 4658-4667.

98. Hébert-Leclerc C. 2011. Study of the immunomodulatory activity of the beta-LG peptide

f96- 99 in healthy mice. Master's thesis, Université Laval, Québec, 104 pages.

99. Hernández-Ledesma B., Hsieh C.-C., de Lumen B.O. Chemopreventive properties of peptide lunasin: a review. *Protein and Peptide Letters,* 2013; 20: 424-432.

100.Hernández-Ledesma B., García-Nebot M. J., Fernández-Tomé S., Amigo L., Recio I. Dairy protein hydrolysates: Peptides for health Benefits.

International Dairy Journal, 2014; 38(2):82-100.

101.Heyman M., Andriantsoa M., Crain-Denoyelle A.M., Desjeux J.F. Effect of oral or parenteral sensitization to cow's milk on mucosal permeability in Guinea pigs. *International Archives of Allergy and Immunology,* 1990; 92:242-246.

102.Heyman M., Boudraa G., Sarrut S., Giraud M., Evans L., Touhami M., Desjeux J.-F. Macromolecular Transport in Jejunal Mucosa of Children with Severe Malnutrition: A Quantitative Study. *Journal of pediatric gastroenterology and nutrition,* 1984; 3: 3, 357-363.

103.Heyman M. Food antigens, the intestinal barrier and mucosal immunity. *Cahiers de nutrition et de diététique,* 2010; 45: 65-71.

104.Holck A., & Naes H. Cloning, sequencing and expression of the gene encoding the cellenvelope-associated proteinase from Lactobacillus paracasei subsp. Paracasei NCDO 151. *Journal of General Microbiology,* 1992; 138: 1353-1364.

105.Host A. Frequency of cow's milk allergy in childhood. *Annals of Allergy, Asthma and Immunology,* 2002 ; 89 :33-6.

106.Hould R. Thecnique d'histopathologie et de cytologie. Montréale. *Décarie,* 1984; 47:156.

107.Ibsaine O., Djenouhat K., Lemdjadani N., Berrah H. Incidence of IgE-mediated cow's milk protein allergy during the first year of life. *Nutrition santé, 2013* ; 2: 9-16.

108.Jacobsen C.N., Rosenfeldt Nielsen V., Hayford A.E., Moller P.L., Michaelsen K.F., Paerregaard A., Sandstrom B., Tvede M., Jakobsen M. Screening of probiotic activities of forty-seven strains of Lactobacillus spp. by in vitro techniques and evaluation of the colonization ability of five selected strains in humans. *Applied and Environmental Microbiology,* 1999; 65: 4949-4956.

109.Jäkälä P., & Vapaatalo H. Antihypertensive peptides from milk proteins. *Pharmaceuticals,* 2010; 3: 251-272.

110.Jankovic I, Sybesma W, Phothirath P, Ananta E, Mercenier A. Application of probiotics in food products-challenges and new approaches. *Current Opinion in Biotechnology.* 2010; 21:175-81.

111.Jedrychowski L., Wroblewskans B. Reduction of the antigenicity if whey proteins by lactic acid fermentation. *Food and Agricultural Immunology,* 1999; 11: 91-99.

112.Jing H., & Kitts D. D. Redox-related cytotoxic responses to different casein glycation products in Caco-2 and int-407 cells. *Journal of Agricultural and Food Chemistry,* 2004; 52: 3577-3582.

113.Joffin J.N. and Leyral G.1996. Microbiologie technique. Centre régional de documentation pédagogique d'Aquitaine, Bordeaux, France: 219-223.

114.Juillard V., Spinnler M., Desmazeaud M. J., Bouquien C.Y. Cooperation and inhibition between lactic bacteria used in the dairy industry. *Lait,* 1987 ; 69: 149-172.

115.Julliard V., Laan H., Kunji E., Jeronimus-stratingh C.M., Bruins A., Konings W. The extracellular PI-type proteinase of lactococcus lactis hydrolyzes β-casein into more than one hundred different oligopeptides. *Journal of Bacteriology,* 1995; 177: 34723478.

116.Kaiserlian D., Lachaux A., Grosjean I., Graber P., Bonnefoy J.Y. Intestinal epithelial cells express the CD23/Fc epsilon RII molecule: enhanced expression in enteropathies.

Immunology, 1993; 80:90-5.

117.Kandler O. Carbohydrate metabolism in lactic acid bacteria. *Antonie Van Leeuwenek,* 1983; 49:209-224

118.Katla A., Kruse H., Johnsen G., Herikstad H. Antimicrobial susceptibility of starter culture bacteria used in Norwegian dairy products. *International Journal of Food Microbiology*, 2001; 67: 147-152.

119.Kawakami T. & Galli S.J. Regulation of mast-cell and basophil function and survival by IgE. *Nature Review of Immunology*, 2002; 2: 773-786.

120.Ke D., Picard F., Martineau F., Menard C., Roy P.H., Oulette M., Bergeron M.G. developpement of *PCR assay for rapid detection of Enterococci. Journal of Clinical Microbiology,* 1999; 37: 3497-3503.

121.Keita A.V., Soderholm J.D. The intestinal barrier and its regulation by neuroimmune factors. *Neurogastroenterology and Motility*, 2010; 22(7):718-733.

122.Khalid N., Marth E. Lactobacilli-their enzymes and role in ripening and spolage of cheese: a review. *Journal of Dairy Science*, 1990; 73: 2669-2684.

123.Kheroua O., Tomé D., Marcon-Genty D., Ben Mansour A., Desjeux J-F. Anticholeraic effect and intestinal transepithelial passage of native and soluble formaldehyde- modified caseins. Padiatric Research, *European Society for Paediatric Research,* 1987; 22(2): 234-234.

124.Kieronczyk A., Skeie S., Olsen K., Langsrud T. Metabolism of amino acids by resting cells of non-starter lactobacilli in relation to flavor development in cheese. *International Dairy Journal*, 2001; 11:217-224.

125.Kleber N., Weyrich U., Hinrichs J. Screening for lactic acid bacteria with potential to reduce antigenic response of β-lactoglobulin in bovine skim milk and sweet whey. *Innovative Food Science and Emerging Technologies*, 2006; 7:233-238.

126.Kmet V., Javorsky P., Nemocova R., Kopecny J., Boda K. Occurrence of conjugative of amylolytic activity in rumen lactobacilli. *Zentralblatt für Mikrobiologie, 1989; 144(1): 53-57.*

127.Koessler K.K., Hanke M.T., Sheppard M.S. Production of histamine, tyramine, brochospastic and arteriospastic substance in blood broth by pure cultures of microorganisms. *Journal of Infectious Diseases,* 1928; 3: 363-377.

128.Korhonen H. and Pihlanto A. Food-derived bioactive peptides opportunities for designing future foods. *Current Pharmaceutical Design*, 2003; 9:1297-1308.

129.Korhonen H., Pihlanto A. Bioactive peptides: Production and functionality. *International Dairy Journal*, 2006; 16, 945-960.

130.Kume H., Okazaki K., Takahashi T., Yamaji T. Protective effect of an immune-modulating diet comprising whey peptides and fermented milk products on indomethacin-induced small-bowel disorders in rats. *Clinical Nutrition,* 2014 ; 33(6):1140-1146.

131.Kunji E. R. S., Mierau I., Hagting A., Poolman B., Konings W.N. Proteolytic systems of lactic acid bacteria. *Antonie van Leeuwenhoek* 1996; 70: 187-221.

132.Laan H., & Konings W.N. Mechanism of proteinase release from Lactococcus lactis subsp. cremoris Wg2. *Applied and Environmental Microbiology*, 1989; 55: 31013106.

133.Lachaux A., Grosjean I., Bonnefoy J.Y., Kaiserlian D. Soluble serum CD23 levels and CD23 molecule expression on intestinal epithelial cells in infants with reaginic and non-reaginic cow's milk allergy. *European Journal of Pediatrics*, 1996; 155:918.

134.Lantz C.S., Yamaguchi M., Oettgen H.C., Katona I.M., Miyajima I., Kinet J.P., and Galli S.J. IgE regulates mouse basophil FceRI expression in vivo. *Journal of Immunology,* 1997; 158: 2517-2521.

135.Larché M., Akdis C., Valenta R. Immunological mechanisms of allergen-specific immunotherapy. *Nature reviews /immunology* volume 6. Nature Publishing Group 2006.Botturi K., Magnan A. Histamine, a new T lymphocyte cytokine? *Revue française d'allergologie et d'immunologie clinique*, 2006 ; 46: 640-647.

136.Larpent J.P. and Larpent M.G. Memento technique de microbiologie. Second Ed technique et documentaire Lavoisier, 1990; 417-420.

137.LeBlanc J. G., Matar C., Valdez J. C., LeBlanc J., & Perdigon G. Immunomodulating effects of peptidic fractions issued from milk fermented with Lactobacillus helveticus. *Journal of Dairy Science*, 2002; 85: 2733- 2742.

138.LeBlanc J., Fliss I., and Matar C. Induction of a humoral immune response following an Escherichia coli O157:H7 infection with an immunomodulatory peptidic fraction derived from Lactobacillus helveticus-fermented milk. *Clinical and Diagnostic Laboratory Immunology*, 2004; 11: 1171-1181.

139.Li H., Sheppard D.N., and Hug M.J. Transepithelial electrical measurements with the Ussing chamber. *Journal of Cystic Fibrosis,* 2004; 3:123-126.

140.Liu D., Sun H., Zhang L., Li S., and Qin Z. High-level expression ofmilk-derived antihypertensive peptide in Escherichia coli and its bioactivity. *Journal of Agricultural and Food Chemistry*, 2007; 55: 5109-5112.

141.López-Expósito R., & Recio I. Antibacterial activity of peptides and folding variants from milk proteins. *International Dairy Journal*, 2006 ; 16: 1294-1305.

142.Lorient D., Closs B., Courthaudon J.L.Connaissances nouvelles sur les propriétés fonctionnelles des protéines du lait et des dérivés. *Lait*, 1991; 71:141-171.

143.Lydyard P.M., Whelan A., Fanger M.W. Essentials of Immunology. Edition Berti: Paris, 2002; 309 p.

144.Maeno M., Yamamoto N., and Takano T. Identification of an antihypertensive peptide from casein hydrolysate produced by a proteinase from Lactobacillus helveticus Cp790. *Journal of Dairy Science*, 1996;79(8): 1316-1321.

145.Mallegol J., van Niel G., Heyman M. Phenotypic and functional characterization of intestinal epithelial exosomes. *Blood Cells, Molecules and Diseases*, 2005; 35: 11-16.

146.Mallegol J., Van N.G., Lebreton C., Lepelletier Y., Candalh C., Dugave C., Heath J.K., Raposo G., Cerf-Bensussan N., Heyman M. T84-intestinal epithelial exosomes bear MHC class II/peptide complexes potentiating antigen presentation by dendritic cells. *Gastroenterology,* 2007; 132:1866-1876.

147.Maragkoudakis P.A., Zoumpopoulou G., Miaris C., Kalantzopoulos G., Pot B., Tsakalidou E. Probiotic potential of Lactobacillus strains isolated from dairy products. *International Dairy Journal*, 2006; 16: 189-199.

148.Marcon-Genty D., Tome D., Kheroua O., Dumontier A. M., Heyman M., Desjeux J.- F. Transport of beta-lactoglobulin across rabbit ileum in vitro. American Journal of Physiology - *Gastrointestinal and Liver Physiology*, 1989; 256: 6, 943-948.

149.Maruyama S, Suzuki H. A peptide inhibitor of angiotensin-I converting enzyme in the tryptic hydrolysate of casein. *Agricultural and Biological Chemistry*, 1982; 46:13931394.

150.Maruyama S., Awaya J., Tomizuka N., Mitachi H., Kurono M., and Suzuki H. Angiotensin I-converting enzyme inhibitor activity of the C-terminal hexapeptide of αs1-casein. *Agricultural and Biological Chemistry,* 1987; 51: 2557-2561.

151.Masco L., Van Hoorde K., De Brandt E., Swings J., Huys G. Antimicrobial susceptibility of Bifidobacterium strains from humans, animals and probiotic products. *Journal of*

Antimicrobial Chemotherapy, 2006; 58(1): 85-94.

152.Matar C., Amiot J., Savoie L., & Goulet J. The effect of milk fermentation by Lactobacillus helveticus on the release of peptides during in vitro digestion. *Journal of Dairy Science*, 1996; 79: 971- 979.

153.Matar C., Valdez J. C., Medina M., Rachid M., & Perdigon G. Immunomodulating effects of milks fermented by Lactobacillus helveticus and its nonproteolytic variant. *Journal of Dairy Research*, 2001; 68: 601- 609.

154.Matar C., LeBlanc J. G., Martin L., Perdigon G. Biologically active peptides released in fermented milk: Role and functions. In E. R. Farnworth (Ed.), Handbook of fermented functional foods. *Functional foods and nutraceuticals series.* 2003, 177-201. Florida, USA: CRC Press.

155.Maxwell S. R. J., and Lip G. Y. H. Free radicals and antioxidants in cardiovascular disease. *British Journal of Clinical Pharmacology*, 1997; 44: 307- 317.

156.McSweeney P., Sousa M. Biochemical pathways for the production of flavor compounds in cheeses during ripening: A review. *Lait*, 2000; 80: 293-324.

157.Meisel H. Biochemical properties of bioactive peptides derived from milk proteins: Potential nutraceuticals for food and pharmaceutical applications. *Livestock Production Science*, 1997; 50: 125- 138.

158.Meisel H., Bockelmann W. Bioactive peptides encrypted in milk proteins: proteolytic activation and thropho-functional properties. *Antonie van Leeuwenhoek*, 1999;76: 207-215.

159.Meisel H., and FitzGerald R. J. Biofunctional peptides from milk proteins: Mineral binding and cytomodulatory effects. *Current Pharmaceutical Design*, 2003; 9: 12891295.

160.Ménard S., Cerf-Bensussan N. and Heyman M. Multiple facets of intestinal permeability and epithelial handling of dietary antigens. Mucosal Immunology, 2010; 3: 247-259.

161.Ménard S., Lebreton C., Schumann M., Matysiak-Bunik T., Dugave C., Bouhnik Y., Malamut G., Cellier C., Allez M., Crenn P., Schuzke J.D., Cerf-Bensussan N., and Hehman M. Paracellular versus transcellular intestinal permeability to gliadin peptides in active celiac disease. *The American Journal of Pathology*, 2012; 180(2):608-615.

162.Mercier A., Gauthier S.F., Fliss I. Immunomodulating effects of whey proteins and their enzymatic digests. *International Dairy Journal* , 2004; 14: 175-183.

163.Merja N., Sirpa J., Marja-Leena L., Hans S., Soili M.-K., Johanna M.K., Nina H., Tari H., Kristiina T., Juha R. Molecular interactions between a recombinant IgE antibody and the β-Lactoglobulin allergen. *Structure,* 2007; 15: 1413-1421.

164.Miclo L., Perrin E., Driou A., Papadopoulos V., Boujrad N., Vanderesse R., Boudier J.F., Desor D., Linden G.,Gaillard J.L. Characterization of α-casozepine, a tryptic peptide from bovine αs1-casein with benzodiazepine-like activity. *Federation of American societies for experimental biology Journal*, 2012; 15: 1780-1782.

165.Migliore-Samour D., Floc'h F., & Jolle's P. Biologically active casein peptides implicated in immunomodulation. *Journal of Dairy Research*, 1989; 56, 357-362.

166.Mikelsaar M. and zilmer M. *Lactobacillus fermentum* ME-3 an ntimicrobial and antioxidative probiotic. *Microbial Ecology in Health and Disease,* 2004; 1:1-27.

167.Minkiewicz P., Slangen C. J., Dziuba J., Visser S., and Mioduszewska H. Identification of peptides obtained via hydrolysis of bovine casein by chymosin using HPLC and mass spectrometer. *Milchwissenschaft*, 2000 ; 55(1) : 14-17.

168.Morali A. Allergies aux protéines du lait de vache e n pédiatrie. *Revue Française des Laboratoires*, 2004 ; 363: 47-55.

169.Moslehishad M., Ehsani M. R., Salami M., Mirdamadi S., Ezzatpanah H., Naslaji A. N., Moosavi-Movahedi A. The comparative assessment of ACE-inhibitory and antioxidant activities of peptide fractions obtained from fermented camel and bovine milk by Lactobacillus rhamnosus PTCC 1637. *International Dairy Journal,* 2013; 29:82-87

170.Muro Urista C., Álvarez Fernández R., Riera Rodriguez F., Arana Cuenca A., & Téllez Jurado A. Review: Production and functionality of active peptides from milk. *Food Science and Technology International*, 2011; 17: 293-317.

171.Murphy MS., walker W.A. Celiac desease. *Pediatrics in Review*, 1991; 12: 325-30.

172.Nakamura Y., Yamamoto N., Sakai K., Okubo A., Yamazaki S., & Takano, T. Purification and characterization of angiotensin-I-converting enzyme inhibitors from sour milk. *Journal of Dairy Science,* 1995; 78(4): 777-783.

173.Negaoui H., Kaddouri H., Kheroua O. & Saidi D. A model of intestinal anaphylaxis in whey sensitized Balb/c mice. *American Journal of Immunology*, 2009; 5:56-60.

174.Negrao-Correa D., Adams L.S., Bell R.G. Intestinal transport and catabolism of IgE: a major blood-independent pathway of IgE dissemination during a Trichinella spiralis infection of rats. *Journal of Immunology*, 1996; 157: 4037-44.

175.Netwich I., Szefalusi Z.,Kunz C., Spurgin P., Urbanek R. Antigenicity for humans of cow milk caseins, casein hydrolysate and casein hydrolysate fractions. *Acta Veterinaria Brno*, 2004; 55:50-61.

176.Netwich I., Szepfalusi Z., Kunz C., Spurgin P., Urbanek R. Antigenicity for humans of cow milk caseins, casein hydrolysate and casein hydrolysate fractions. *Acta Veterinaria Scandinavica*, 2004; 73:291-298.

177.Panesar P. S. Fermented Dairy Products: Starter Cultures and Potential Nutritional Benefits. *Food and Nutrition Sciences*, 2011; 2: 47-51

178.Pastar I., Tonic I., Golic N., Kojic M., van Kranenburg R., Kleerebezem M., Topisirovic L., Jovanovic G. Identification and genetic characterization of a novel proteinase, PrtR, from the human isolate Lactobacillus rhamnosus BGT10. *Applied and Environmental Microbiology*, 2003; 69, 5802-5811.

179.Pederson J., Mileski G., Weimer B., Steele J. Genetic characterization of a cell envelope associated proteinase from lactobacillus helveticus CNRZ32. *Journal of Bacteriology*, 1999 ; 181: 4592-4597.

180.Pescuma M., Hébert E., Mozzi F., Font de Valez G. Whey fermentation by thermophilic lactic acid bacteria: Evolution of carbohydrates and protein content. *Food Microbiology,* 2008, 25: 442-451.

181.Pescuma M., Hébert E. M., Rabesona H., Drouet M., Choiset Y., Haertlé T., Mozzi F., Font de Valdez G., Chobert J.-M. Proteolytic action of Lactobacillus delbrueckii subsp. bulgaricus CRL 656 reduces antigenic response to bovine b-lactoglobulin. *Food Chemistry*, 2011 ; 127: 487-492.

182.Phelan M., Aherne A., Fitzgerald R. J., & O'Brien N. M. Casein-derived bioactive peptides: biological effects, industrial uses, safety aspects and regulatory status. *International Dairy Journal*, 2009b; 19: 643-654.

183.Pihlanto A. Antioxidative peptides derived from milk proteins. *International Dairy Journal*, 2006; 16: 1306- 1314.

184.Powell D.W. Barrier functions of epithelia. *American Journal of Physiology,* 1981; 241: 275-288.

185.Qian B., Xing M., Cui L., Deng Y., Xu Y., Huang M., Zhang S. Antioxidant, antihypertensive, and immunomodulatory activities of peptide fractions from fermented skim milk with *Lactobacillus delbrueckii ssp. bulgaricus* LB340. *Journal of Dairy Research*, 2011; 78: 72-79.

186.Rahimi E., M. Ameri and H. R. Kazemeini. Prevalence and antimicrobial resistance of Campylobacter species isolated from raw camel, beef, lamb, and goat meat in Iran. *Foodborne Pathogens and Disease*, 2010 ; 7:443-447.

187.Rance F., Lymphocytes T et allergies alimentaires. *Revue française d'allergologie et d'immunologie clinique*, 2007; 47: 214-218.

188.Raposo G., Nijman H.W., Stoorvogel W., Liejendekker R., Harding C.V., Melief C.J., and Geuze H.J. B lymphocytes secrete antigen-presenting vesicles. *Journal of Experimental Medicine*, 1996; 183: 1161-1172.

189.Re R., Pellegrini N., Proteggente A., Pannala A., Yang M., Rice-Evans C., Antioxidant activity applying an improved ABTS radical cation decolorization assay. *Free Radical Biology and Medicine*, 1999; 26: 1231-1237.

190.Rea M., Cogan T. Glucose prevents citrate metabolism by enterococci. *International Journal of Food Microbiology*, 2003; 88N 2/3: 201-206.

191.Regazzo D., Da Dalt L., Lombardi A., Andrighetto C., Negro A., & Gabai G. Fermented milks from Enterococcus faecalis TH563 and Lactobacillus delbrueckii subsp. bulgaricus LA2manifest different degrees of ACE-inhibitory and immunomodulatory activities. *Dairy Science and Technology,* 2010; 90: 469-476.

192.Rival S. G., Fornaroli S., Boeriu C. G., and Wichers H. J. Caseins and casein hydrolysates. 1. Lipoxygenase inhibitory properties. *Journal of Agricultural and Food Chemistry*, 2001; 49: 287-294.

193.Ross G., Louise E. Cow's Milk Allergy: A Complex Disorder. *Journal of the American College of Nutrition*, 2005; 24:582-591.

194.Ruttarattanamongkol K. Functionalization ofwhey proteins by reactive supercritical fluid extrusion. *Songklanakarin Journal of Science and Technology*, 2012; 34, 395-402.

195.Saidi D., Heyman M., Kheroua O., Boudraa G., Bylsma P., Kerroucha R., Chekroun A., Maragi J.-A., Touhami M., Desjeux J.-F. Jejunal response to β-lactoglobulin in infants with cow's milk allergy. Proceedings of the French Academy of Sciences. Series 3, *Life Sciences*, 1995; 318(6): 683-689.

196.Salami M., Yousefi R., Ehsani M. R., Razavi S. H., Chobert J.M., Haertle T., Saboury A. A., Atri M. S., Niasari-Naslaji A., Ahmade F., Moosavi-Movahedi A. A. Enzymatic digestion and antioxidant activity of the native and molten globule states of camel a- lactalbumin: Possible significance for use in infant formula. *International Dairy Journal,* 2009; 19: 518-523.

197.Sandré C., Gleizes A., Forestier F., Gorges-Kergot R., Chilmonczyk S., Leonil J., Moreau MC, Labarre C. A peptide derived from bovine beta-casein modulates functional properties of bone marrow-derived macrophages from germfree and human flora-associated mice. *Journal of Nutrition*, 2001 ; 131: 2936-2942

198.Santos A.C. Where and how to induce tolerance in milk-allergic children. *Revue française d'allergologie,* 2014 ; 54 :183-187.

199.Sarantinopoulos P., Andrighetto C., Georgalaki M. D., Rea M. C., Lombardi A., Cogan T. M., Kalantzopoulos G., Tsakalidou E. Biochemical properties of enterococci relevant to their technological performance. *International Dairy Journal*, 2001; 11: 621-647.

200.Sarantinopoulos P., Kalantzopoulos G., Tsakalidou E. Citrate metabolismby

Enterococcus faecalis FAIR-E229. *Applied and Environmental Microbiology*, 2001; 67:5482-5487.

201.Savijoki K., Ingmer H., & Varmanen P. Proteolytic systems of lactic acid bacteria. *Applied Microbiology and Biotechnology*, 2006; 71: 394-406.

202.Schagger H., Aquila H., AND Vonjagow G. Coomassie Blue-Sodium Dodecyl SulfatePolyacrylamide Gel Electrophoresis for Direct Visualization of Polypeptides during Electrophoresis. *Analytical Biochemistry*, 1988; 173(20): 1-205.

203.Schillinger U. and Lucke F. Antimicrobial activity of Lactobacillus sake isolated from meat. *Applied and Environmental Microbiology*, 1989; 55:1901-1906.

204.Schleifer K.H., Ludwig W. Phylogeny of the genus Lactobacillus and related genera. *Systematic and Applied Microbiology*, 1995b; 18: 461-467.

205.Schwan H.P. Electrode polarization impedance and measurements in biological materials. *Annual New York Academy of Science*, 1968; 148(1):191-209.

206.Selo I., Negroni L., Creminon C., Yvon M., Peltre G., and Wal J.M. Allergy to bovine β-lactoglobulin: specificity of human IgE using cyanogen bromide- derived peptides. *International Archive of Allergy and Immunology*, 1998; 117: 20-28.

207.Sharma S., Kumar P., Betzel C., Singh T.P. Structure and function of proteins involved in milk allergies. *Journal of Chromatography B*, 2001; 756:183-187.

208.Shori A. B. Antioxidant activity and viability of lactic acid bacteria insoybean-yogurt made from cow and camel milk. *Journal of Taibah University for Science*, 2013; 7: 202-208.

209.Sicherer S.H., Sampson H.A. Cow's milk protein-specific IgE concentrations in two age groups of milk-allergic children and in children achieving clinical tolerance. *Clinical and Experimental Allergy*, 1999; 29(4):507-512.

210.Siezen R. J. Multi-domain, cell-envelope proteinases of lactic acid bacteria. *Antonie Van Leeuwenhoek*, 1999; 76, 139-155.

211.Silva S. V., Malcata F. X. Caseins as source of bioactive peptides. *International Dairy Journal*, 2005; 15:1-15.

212.Sisto A., and Lavermicocca P. Suitability of a probiotic *Lactobacillus paracasei* strain as a starter culture in olive fermentation and development of the innovative- patented product "probiotic table olives". *Frontiers in Microbiology,* 2012; 3.

213.Skripak J. M., Matsui E. C., Mudd K., Wood R. A. The natural history of IgE-mediated cow's milk allergy. *Journal of Allergy and Clinical Immunology*, 2007; 120:11721177.

214.Sorenson R.U., Porch M.C. and Tu L.C. Food allergy in children. Text book of pediatric Nutrition 2nd edition, edited by RM skind and Lewinter'Suskind. New York Raven Press, ltd, 1993; 457-469.

215.Spahn T.W., Kucharzik T. Modulating the intestinal immune system: the role of lymphotoxin and GALT organs.*Gut,* 2004; 53(3):456-465.

216.Stackebrandt E., Teuber M. Molecular taxonomy and phylogenetic position of lactic acid bacteria*, Biochimie*, 1988; 70 (3):317-24.

217.Stanley G., Schultz and Zalusky R. Ion Transport in Isolated Rabbit Ileum. *The Journal of General Physiology,* 1964; 47(3): 567-584.

218.Stefanitsi D., Sakellaris G., Garel J. The presence of two proteinases associated with the cell wall of *lactobacillus bulgaricus*. *FEMS Microbiology Letters,* 1995; 128:53-58.

219.Strahinic I., Kojic M., Tolinacki M., Fira D., Topisirovic L. The presence of prtP Proteinase gene in natural isolate *Lactobacillus plantarum* BGSJ3-18. *Letters in Applied Microbiology,* 2010; 50: 43-49.

220.Suetsuna K., Ukeda H., and Ochi H. Isolation and characterization of free radical scavenging activities peptides derived from casein. *Journal of Nutrition and Biochemistry*, 2000; 11, 128-131.

221.Sütas Y, Hurme M, Isolauri E. Down-regulation of anti-CD3 antibody-induced IL-4 production by bovine caseins hydrolysed with Lactobacillus GG-derived enzymes. *Scandinavian Journal of Immunology*, 1996; 43: 687-689.

222.Suzzi G., Caruso M., Gardini F., Lombardi A., Vannini L., Guerzoni M. E., Andrighetto C., & Lanorte M. T. A survey of the enterococci isolated from an artisanal Italian goat's cheese (semicotto caprino). *Journal of Applied Microbiology*, 2000; 89: 267-274.

223.Takano T. Anti-hypertensive activity of fermented dairy products containing biogenic peptides. *Antonie van Leeuwenhoek, 2002; 82: 333-340.*

224.*Tauzin J., Miclo L., and Gaillard J.-L. Angiotensin-Iconverting enzyme inhibitory peptides from tryptic hydrolysate of bovine as2-casein. FEBS Letters, 2002; 531(2):* 369-374.

225.*Terzaghi B.E. and sandine W.E. Improved mediumfor lactic streptococci and their bacteriophages. Journal of Applied Microbiology, 1975; 29: 807-813.*

226.*Tsakalidou E., Manolopoulou E. Tsilibari V., Georgalaki M., Kalantzopoulos G. Esterolytic activities of Enterococcus durans and Enterococcus faecium strains isolated from Grrk cheese. Ntherlands Milk and Dairy Journal, 1993; 47:145-150.*

227.*Tsuge I., Kondo Y., Tokuda R., Kakami M., Kawamura M., Nakajima Y., Komatsubara R., Yamada K., Urisu A. Allergen-specific helper T cell response in* patients with cow's milk allergy: simultaneous analysis of proliferation and *cytokine production by. Clinical and Experimental Allergy, 2006; 36: 1538-1545.*

228.*Tzvetkova I., Dalgalarrondo M., Danova S., Iliev I., Ivanova I., Chobert J.-M., Haertlé T. Hydrolysis of major dairy proteins by lactic acid bacteria from Bulgarian yogurts. Journal of Food Biochemistry, 2007; 31: 680-702.*

229.*Ussing Hans H., Zerahn K. Active Transport of Sodium as the Source of Electric Current in the Short-circuited Isolated Frog Skin. Acta Physiological Scandinavica, 1951; 23, (2-3):* 110-127.

230.Van Niel G., Raposo G., Candalh C., Boussac M., Hershberg R., Cerf-Bensussan N., Heyman M. Intestinal epithelial cells secrete exosome-like vesicles. *Gastroenterology*, 2001; 121: 337-349.

231.Veljovic K., Fira D. Terzic-Vidojevic A., Abriouel H. Galvez A., Topisirovic L. Evaluation of antimicrobial and proteolytic activity of Enterococci isolated from fermented products. *European Food Research and Technology*, 2009; 230: 63-70.

232.Vermeirssen V., Van Camp J., & Verstraete W. Bioavailability of angiotensin I converting enzyme inhibitory peptides. Review *British Journal of Nutrition,* 2004; 92: 357-366.

233.Villani F., & Coppola S. Selection of enterococcal strains for water-buffalo Mozzarella cheese manufacture. *Annales Microbiologia Enzimologia,* 1994; 44: 97-105.

234.Vinderola G., Matar C., Palacios J., Perdigón G. Mucosal immunomodulation by the non-bacterial fraction of milk fermented by *Lactobacillus helveticus* R389. *International Journal of Food Microbiology*, 2007; 115: 180-186.

235.Wang W., De Mejia E. G. A new frontier in soy bioactive peptides that may prevent age-related chronic diseases. *Comprehensive Reviews in Food Science and Food Safety,* 2005; *4:63-78.*

236.Weisburg W., Barns S., Pelletier D., Lane D.16S ribosomal DNA amplification for phylogenetic study. *Journal of Bacteriology*, 1991; 137: 697-703.

237.Williams A., Withers S., Banks J. Energy sources of non-starter lactic acid bacteria isolated from Cheddar cheese. *International Dairy Journal*, 2000; 10: 17-23.

238.Wood B., Holzapfel W., Banks J. The genera of lactic acid bacteria, 1st ed. *Blackie Academic and professional, Glasgow*, United Kingdom, 1995; p.42.

239.Wood B., Warner P. Genetics of lactic acid bacteria. *Kluwer Academic.Plenum Publshers*, New Yorrk, 2003, p. 26.

240.Wrobleswska B., Jedrychowski L., Bielecka M. The effect of selected microorganisms on the presence of immunoreactive fractions in cow and goat milks. Polish *Journal of Nutrition & Food Sciences,* 1995; 4/45(3):21-28.

241.Xanthopoulos V., Litopoulou-Tzanetaki E., Tzanetakis N. Characterization of Lactobacillus isolates from infant faeces as dietary adjuncts. *Food Microbiology,* 2000; 17(2): 205-215.

242.Xiong Y. L. Antioxidant peptides. In Y. Mine, B. Jiang, & E. Li-Chan (Eds.), Bioactive proteins and peptides as functional foods and neutraceuticals (pp. 29^e 39). Ames, Iowa, USA (2010): Wiley-Blackwell.

243.Yamamoto Y., Togawa Y., Shimoska M., Okazaki M. Purification and characterization of novel bacteriocil produced by *Enterococcus faecalis* strain Rj-11. *Applied and Environmental Microbiology*, 2003; 69: 5746- 5753.

244.Yang P.C., Berin M.C., Yu L.C., Conrad D.H., Perdue M.H. Enhanced intestinal transepithelial antigen transport in allergic rats is mediated by IgE and CD23 (Fcepsilon RII). *Journal of Clinical Investigation*, 2000; 106: 879-86.

245.Zacarías M. F., Binetti A., Laco M., Reinheimer J., and Vinderola G. Preliminary technological and probiotic characterization of bifidobacteria isolated from breast milk for use in dairy products. *International Dairy Journal*, 2011 ; 21:548-555.

246.Zellal D. 2009. Effect of the combined treatment of microwave heating and pH on the allegenicity of cow's milk proteins in Balb/c mice.

247.Zellal D., Kaddouri H., Grar H., Belarbi H., Kheroua O. and Saidi D. Allergenic changes in β-lactoglobulin induced by microwave irradiation under different pH conditions. *Food and Agricultural Immunology*, 2011; 22: 355-363.

248.Zhao S., P. F. McDermott S. Friedman J. Abbott S. Ayers A. Glenn E. Hall-Robinson S. K. Hubert H. Harbottle R. D. Walker T. M. Chiller and D. G. White. Antimicrobial resistance and genetic relatedness among Salmonella from retail foods of animal origin: NARMS retail meat surveillance. *Foodborne Pathogens and Disease,* 2006; 3: 106-117.

249.Zhao X. H., Wu D., Li T. J. Preparation and radical scavenging activity of papain-catalyzed casein plasteins. *Dairy Science Technology*, 2010; 90: 521-535.

250.Zhou N., Zhang J. X., Fan M. T., Wang J., Guo G., and Wei X. Y. Antibiotic resistance of lactic acid bacteria isolated from Chinese yogurts. *Journal of Dairy Science*, 2011 ; 95:4775-4783.

More
Books!

info@omniscriptum.com
www.omniscriptum.com
OMNIScriptum

Printed by Books on Demand GmbH, Norderstedt / Germany